Biosystems & Biorobotics

2

Series Editor

Eugenio Guglielmelli
Laboratory of Biomedical Robotics & Biomicrosystems
Università Campus Bio-Medico
Roma
Italy
E-mail: E.Guglielmelli@unicampus.it

For further volumes:
http://www.springer.com/series/10421

Ralf Simon King

BiLBIQ: A Biologically Inspired Robot with Walking and Rolling Locomotion

 Springer

Author
Ralf Simon King M.Sc.
Fraunhofer IPA
Stuttgart

ISSN 2195-3562   e-ISSN 2195-3570
ISBN 978-3-642-44846-1   ISBN 978-3-642-34682-8 (eBook)
DOI 10.1007/978-3-642-34682-8
Springer Heidelberg New York Dordrecht London

Foreword

Ein kurzes Vorwort

Vorweggenommen: Es ist spannend, die Arbeit von Ralf Simon King zu lesen. Eine Spinne, so haben wir es in der Schule gelernt, läuft auf 8 Beinen. So sollte es auch ein Spinnenroboter machen – denkt man. Aber seit Neuestem weiß man, dass eine Saharaspinne der Gattung *Cebrennus* auch mit ihren 8 Beinen rollen kann. Ein neues Feld der Bio-Robotik tut sich auf.

Cebrennus kann achtbeinig krabbeln und Purzelbaum schlagend rollen, so wie es die Situation erfordert. Es gibt viele Vorschläge, auch Roboter zu konstruieren, die laufen und rollen können. Mit der Saharaspinne wird die Lösung der biologischen Evolution favorisiert. Ralf Simon King analysiert anhand von Videoaufnahmen eingehend die Beinbewegung der rollenden *Cebrennus*. Sein Hardware-Ergebnis, ein Quadroped-Roboter, der wie *Cebrennus* laufen und rollen kann, fasziniert.

Ein Mars-Rover nach dem Vorbild der rollenden Saharaspinne hat einen Nachteil: Alles dreht sich, es gibt keine raumfeste Plattform. Ich denke mir, dass der *Cebrennus*-Rover eher als laufender und rollender Kundschafter (Scout) Anwendung finden könnte. Der 6-rädrige Mars-Rover "Spirit" hat sich im Sand festgefahren. Dem Mars-Rover Curiosity wird es möglicherweise ähnlich ergehen, denn Räder schieben den Sand nach hinten und wühlen sich so ein. Sich rollend fortzubewegen, indem man sich mehr senkrecht vom Boden abstößt, ist im losen Dünensand die bessere Lösung. Und wird das Gelände rau mit vielen Hindernissen, beginnt der Cebrennus-Rover auf seinen Beinen zu laufen.

Ich wünsche der hochaktuellen und mit Genuss lesbaren Publikation von Ralf Simon King viel Erfolg und eine weite Verbreitung.

Ingo Rechenberg

A Short Foreword

Let me anticipate this: It is going to be exciting to read Ralf Simon King's book.

A spider walks on eight legs – that's what we learned in school. Consequently it is what we expect of a spider robot too. However, for a short while now, we have known that a Sahara spider of the species *Cebrennus* can also roll with the use of its eight legs. This led to the opening of a new field in bio-robotics. By simultaneously using all legs available *Cebrennus* is - depending on what the actual situation requires - capable of either scuttling or rolling in a somersaulting manner. There are also many suggestions to construct robots that can both walk and roll. The preferred solution is the one provided by biological evolution: the Sahara spider. Ralf Simon King analyses the leg movement of the rolling *Cebrennus* in great detail with the help of video footage. The result, a quadruped robot that can both walk and roll like the *Cebrennus,* is extremely fascinating.

A Mars rover modelled on the rolling Sahara spider has one disadvantage: everything is spinning; there is no fixed platform. I believe that the *Cebrennus* rover could operate better as a walking and rolling scout. The six-wheeled Mars rover Spirit got stuck in the sand. The Mars rover Curiosity is probably going to suffer the same fate since wheels shovel the sand towards the rear and thus tend to burrow. The better solution, when dealing with loose dune sand, is to make way in a rolling locomotion by pushing oneself off the ground in a vertical direction. In addition, when the terrain gets rough, whereby there are many obstacles, the *Cebrennus* rover could simply start to run on its legs.

I hope the publication by Ralf Simon King, which not only discusses a highly topical issue but is also a pleasure to read, is received with great interest and spread widely.

Ingo Rechenberg

Acknowledgment

I gratefully acknowledge the support that I have received from Prof. Ingo Rechenberg from TU Berlin on whose research this book is actually based upon. I want to thank him for the permission to use both his research data and media for my book. Furthermore, I want to thank him for the open and creative exchange of ideas and knowledge in our interviews about the *Cebrennus villosus* and the potential robotic implementations this unique animal holds.

I also gratefully acknowledge the supervision that I have received from Prof. Thorsten Brandt and Prof. Peter Kisters during my masters' thesis that constitutes the foundation of this book. A special thank goes to Dr. Leontina Di Cecco from Springer Verlag for her kind and patient support that finally made this publication possible. Moreover, I want to thank Adam Braun, a former fellow student of mine, for the fact that we got started together in our initial research with so much eagerness. This is what eventually sparked my interest in this yet existing field of research. The favorable results we could obtain from our project, however, were an even more determining factor that helped promote my idea to write this book and are thus what I explicitly want to thank him for.

A special thank goes to Simon Weniger for the time he spend proofreading my text and suggesting grammatical enhancements for it. Other thanks go to Hendrik van Leeuwen and Maurice Spitzer for their support on building the robot prototypes in the university's workshop. Last but not least a precious thank goes to my parents Gabriele Escher-King and Nikolaus King who always supported me with an encouraging word and a comforting thought.

Contents

List of Figures

List of Tables

List of Abbreviations

°	degree	
3D	Three Dimensional	
ALDURO	Anthropomorphically Legged and Wheeled Duisburg Robot	
ASR	active-spoke-wheel	[Aktivspeichenrad]
BiLBIQ	Bi-Locomotional Biomimetically Inspired Quadruped	
BioKoN	Bionics Competence Network	[Bionik-Kompetenz-Netz]
CAD	Computer Aided Design	
DOF	Degree of Freedom	
EAP	Electroactive Polymer	
FEM	Finite Element Modeling	
g	gram	
IPA	Institute for Manufacturing Engineering and Automation	[Institut für Produktionstechnik und Automatisierung]
JPL	Jet Propulsion Laboratory	
m	meter	
m/s	meter per second	
ml	milliliter	
mm	millimeter	
ms	millisecond	
NASA	National Aeronautics and Space Administration	
RC	Radio Controlled	
SDM	Shape Deposition Manufacturing	
SLS	Selective Laser Sintering	
SMA	Shape Memory Alloy	
SME	Shape Memory Effect	
SpiderBall	Project name of the robot from the previous project	
TU	Technical University	[Technische Universität]

1 Introduction

1.1 Preamble

This book is derived from my master's thesis through which I have achieved the degree 'Master of Science' in the study course Bionics/Biomimetics. The basic idea for it came up during a semester innovation project in 2011. Back then, my fellow student Adam Braun and I were inspired by a spider that is capable of actively doing somersaults using its legs in a very specific way. The question we asked ourselves was: Is it possible to mimic this rolling and walking locomotion with an analog robotic structure? All spiders share the capability of walking, running and jumping, for instance. But seeing a spider form an abstract three-dimensional sphere to roll seemed quite intriguing to us. A robot kit that was used to build a quadruped and a normal basketball from a toy store were our initial research equipment .The basketball served as a mold for spherically shaped feet, which were later attached to the quadruped as shown in the figure below. At the end of the project we were able to let the robot crawl over the floor and let it tumble in a periodic movement pattern. The structure design and hardware challenges, such as exceeding the maximum torque the servo motors could perform, however, avoided a proper walking gait. The robot also was not able to roll properly first since the periodic movement pattern of the legs led to a tumbling motion that would start and stop without control. Nevertheless, the results were promising and a good starting point for further research.

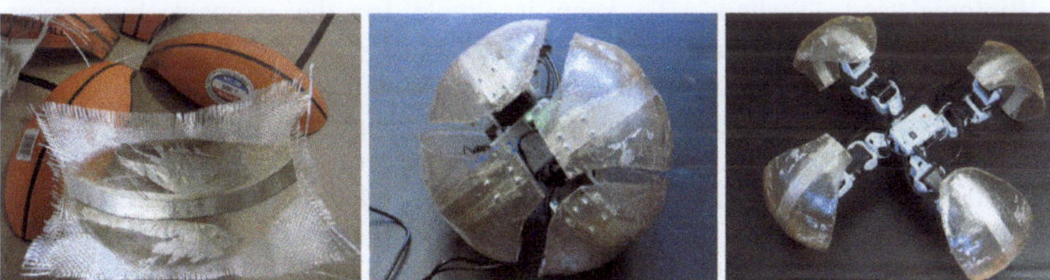

Fig. 1.1 *(left)* Basketball quarters as molds for aluminum and two-component-adhesive reinforced glass-fiber legs. *(middle)* Robot in spherical shape ready for tumbling. *(right)* Robot unfolded ready for crawling.

Ralf Simon King: BiLBIQ: A Biologically Inspired Robot, BIOSYSROB 2, pp. 1–6.
DOI: 10.1007/ 978-3-642-34682-8_1   © Springer-Verlag Berlin Heidelberg 2013

1.2 Tenor Questions

The rolling quadruped robot from the innovation project was neither able to walk
properly nor to roll continuously. Hence the following question had to be asked: Is
it possible to mimic two fundamentally different ways of locomotion with one and
the same structure - **effectively**? Is this this merged structure going to be a
combination of the advantages, or will it just combine its disadvantages? Is it
possible to make difficult terrain accessible by walking and combine this with a
rolling locomotion for plain surfaces? Or will a structure designed for rolling
make proper walking impossible? How effective, in terms of locomotion
performance, will a bio-inspired implementation be, compared to its biological
archetype? Both research on state of the art examples and building the robot for
experimental purposes is expected to give answers to these questions. Therefore a
suitable design of the robot's structure needs to be developed. Mechatronic limits
of given hardware actuators need to be taken into account. Material choice and
usage will be important for a well controllable locomotion of the robot.

1.3 Terms and Definitions

Even though the idea of taking a look at the creations of nature through evolution
and natural selection is not novel, the concept of merging the fields of biology and
engineering is not yet part of general knowledge. The existence of various
definitions exacerbates finding a consistent term for this field of science. In fact,
the possibilities of any kind of bio-related research are enormous. Hence a brief
look shall be taken on some terms that might be misunderstood by people of
different scientific backgrounds.

1.3.1 Bionics

From a scientific point of view bionics and biomimetics are understood equally as
"artificial concepts. Bionics is a made-up word derived from biology and
technics."[1] Likewise, the German word 'Bionik' is formed by the words
'Biologie' and 'Technik'. To keep it simple for the moment, bionics and
biomimetics can be defined as a concept to transfer or apply knowledge gained
from nature into or on technology.

By taking a look beyond scientists and other people working on topics related
to the field of bionics, it can be observed that the term bionics is also understood
in a different way, especially throughout the UK and US. Bionics is, among other
definitions, described as "having parts of the body that are electronic, and
therefore able to do things that are not possible for normal humans"[2]. Such an
artificial body part could be a limb, replacing a lost leg or arm for example that
then could be a bionic amputation appliance. But the definition goes beyond this

---

[1] Gleich, Pade, Petschow, Pissarskoi (2010) p 10.
[2] Oxford Advanced Learner's Dictionary (2010) p 139.

idea and suggests some sort of an electronic 'super' device. This misconception might be explained by a quick look on the early bionics' history.

The first bionic wave in the 60s and 70s of the last century ended up quite early in disappointment when it became clear that a television tower cannot be easily built like a blade of grass. While scientists in Germany were looking for more fundamental approaches, bionics research in the former Soviet Union focused more on military applications. Fantasy filmmakers from the US produced blockbusters like 'Bionic Woman' and 'RoboCop'. As a result, English scientists tried to differentiate the scientific disciplines by renaming the one from bionics into biomimetics.[3]

Having this knowledge another composition of the word bionics can be explained, consisting "of the two words *biology* and *electronics*."[4] In fact there is not just a difference in the meaning of the terms bionics and biomimetics itself, depending on the point of view taken. There is also the "definitions suggested range from simple inspiration to the most exact copy possible."[5]

1.3.2 Biomimetics

From a German point of view the term biomimetics seems not to be entirely well chosen.[6] Merging the word biology with mimetic, "(technical or formal) copying the behaviour or appearance of sb/sth else"[7], works against the definition of extracting and understanding principle functions in nature and applying them on technical solutions.

Nevertheless, the term bionics is biased with thoughts about 'super humans', at least in the average cognitive surroundings of English people. Further definitions of related fields to this book are based on the term biomimetics and hence this is the term that comes to use from here on, by simultaneously covering the term bionics.

1.3.3 Bio-inspired

A closer look at some biomimetic applications might uncover a further misconception. It is necessary to differentiate between biomimetics as a scientific approach and bio-inspired structures and applications that do not even copy a biological solution. Good examples for bio-inspired but not biomimetic designs can be found in architecture. "*Victoria amazonica* leaves were supposedly the inspiration for the greenhouse roof designed by Joseph Paxton (1803-65) and the roof of the Crystal Palace in London, UK, but the geometry and structural

---

[3] cf Bannasch (2002).

[4] URL: `http://www.biokon.net/bionik/bionik.html.en`.
Accessed 27 Dec 2011.

[5] Gleich, Pade, Petschow, Pissarskoi (2010) p 14.

[6] cf Bannasch (2002).

[7] Oxford Advanced Learner's Dictionary (2010) p 973.

principles of the underside of the leaf [...] and that of the roofs are totally different."[8] Inspiration in terms of shape and appearance has barely anything to do with biomimetics and natural principles might be only obeyed by coincidence.

In some fields of research it is not possible, or necessary, to completely understand and transfer functions from nature into applications. But nature still can be a source of inspiration. This applies for artificial intelligence, for instance. Problem solving methods used by living beings can only be understood to a certain degree, but sometimes it is not even necessary to entirely understand the biological archetype. Hence, main principle functions can be extracted, abstracted and applied to software applications, for example. At this point borders between biomimetics and bio-inspired applications become blurry and show how wide these fields of research range.

1.3.4 Biotechnology

When principles of biology are applied on technology, in a wide sense, it can be called biomimetics. At a first glance this needs to be differentiated from the term biotechnology which is in fact: "The application of science and technology to living organisms, as well as parts, products and models thereof, to alter living or non-living materials for the production of knowledge, goods and services."[9] But a closer look at biotechnologies uncovers a highly branched field of bio-inspired and bio-interacting research.

"In its widest definition, biotechnology is all application areas where biology is applied for technical or methodological purposes."[10] A narrower example would be the field of nanobiotechnology and nanotechnology as subgenres of biotechnology. The capability "to manipulate atoms and molecules within materials one at a time, and therefore to construct materials with nanoscale [...] precision [...] is called nanotechnology."[11] Nanotechnology plays a role as a tool for biomimetic applications. Surface structures mimicking properties of biological samples are relying highly on nanoscale effects such as the Lotus-Effect[12]. Nano-biotechnology as a subgenre of nanotechnology could be understood as the biomimetic approach in terms of the Bio-to-nano-approach, when following the rules, construction plans and order principles of nature.[13] Here again the term mimetic leads to a misconception, because nanobiomimetics is not meant to exactly copy a sample given. Instead of mimicking properties, the active process of bionics is to be understood and might thus be called nanobionics.[14] When

---

[8] Allen (2010) p 165.

[9] URL: `http://www.oecd.org/document/42/`
`0,3746,en_2649_34537_1933994_1_1_1_1,00.html`. Accessed 27 Dec 2011.

[10] Mandenius, Björkman (2011) p 42.

[11] Saltzman (2009) p 492.

[12] cf Barthlott, Neinhuis (1997) p 5.

[13] cf Hartmann (2006) p 43 ff.

[14] cf Nachtigall (2002) p 124.

talking about biotechnology, a relation to bioengineering is quickly discovered. Again, at the first glance, it seems to be far away from the field of biomimetics when systems for "a biochemical reaction, a biomedical device, a genetic construct, or a bioanalytical instrument"[15] are engineered and designed. But among other key features bioengineering also means to rebuild functions of human organs with technical means like "artificial liver devices and artificial heart veins"[16], for example. Looking back at the range of definitions for biomimetics and bioengineering, with the mentioned key feature of rebuilding human organs with technical means, it could be considered as true biomimetics since the purpose is to exactly understand and copy biological samples.

1.3.5 Classification of This Book

In order to verify and estimate this book itself and its relevance to biomimetics, it needs both to be classified and linked to the right categories within this field. There are multiple approaches for classification of biomimetics, but a rough subdivision into three main strands is possible:[17]

- Form and Function (Morphology)
- Biocybernetics, Sensor Technology and Robotics
- Nanobiomimetics

The main part of this document is clearly to be attributed to the class of robotics that can also be found on the classifications listed on the BioKoN website, which distinguishes seven classes. Here class B5 for "Biomechatronics, biomedical technology, microelectromechanical systems (MEMS), actuatorics, robotics"[18] is the accurate one. Robotics in biomimetics can still be subdivided into several disciplines. Referring to Nachtigall, the class "Robotics and Locomotion"[19] is the most precise subdivided class among biomimetics, describing exactly the topic of this book, even though rolling as locomotion is not mentioned explicitly in his text.

Furthermore, form and function play a role in this book. Thereby the principles of 'good' natural structure design should be obeyed. But the results of this are of less importance, due to the lack of evidential calculations for the designed structures. Materials are only relevant for a tough prototype without relation to biomaterials. The robot's program is written to demonstrate its functionality purposes only and does not contain any 'artificial intelligence' such as problem solving, optimization or reconfiguration.

---

[15] Mandenius, Björkman (2011) p 42.

[16] Mandenius, Björkman (2011) p 42.

[17] Gleich, Pade, Petschow, Pissarskoi (2010) p 20, Figure 0.

[18] URL: `http://www.biokon.net/bionik/beispiele.html.en`. Accessed 28 Dec 2011.

[19] cf Nachtigall (2002) p 175.

References

Allen, R.: Bulletproof feathers: how science uses nature's secrets to design cutting-edge technology. Ivy Press, Chicago (2010)

Bannasch, R.: Forschen, analysieren, gestalten. Design report 9/2002, pp. 20–25 (2002)

Barthlott, W., Neinhuis, C.: Purity of the sacred lotus, or escape from contamination in biological surfaces. Planta 202(1), 1–8 (1997),
`http://www.springerlink.com/content/4m4jnt03qlvaywjl/`
`fulltext.pdf`, doi:10.1007/s004250050096 (accessed December 28, 2011)

von Gleich, A., Pade, C., Petschow, U., Pissarskoi, E.: Potentials and Trends in Biomimetics. Springer, Heidelberg (2010)

Hartmann, U.: Nanotechnologie. Spektrum Akademischer Verlag, München (2006)

Mandenius, C.-F., Björkman, M.: Biomechatronic Design in Biotechnology: A Methodology for Development of Biotechnological Products. John Wiley & Sons, Hoboken (2011)

Nachtigall, W.: Bionik: Grundlagen und Beispiele für Ingenieure und Naturwissenschaftler, 2nd edn. Springer, Heidelberg (2002)

Oxford Advanced Learner's Dictionary, 8th edn. Oxford Univ. Press, Oxford (2010)

Saltzman, W.M.: Biomedical Engineering: Bridging Medicine and Technology. Cambridge University Press, Cambridge (2009)

`http://www.biokon.net/bionik/bionik.html.en`
(accessed December 27, 2011)

`http://www.oecd.org/document/42/0,3746,en_2649_34537_1933994_`
`1_1_1_1,00.html` (accessed December 27, 2011)

`http://www.biokon.net/bionik/beispiele.html.en`
(accessed December 28, 2011)

2 Biological Archetypes and Robotic Pendants

2.1 The Huntsman Spider *'Cebrennus villosus'*

On January 27th in 2012 I was able to meet Professor Ingo Rechenberg from TU Berlin. According to all information available to me he is the one who discovered the rolling locomotion in the huntsman spider *Cebrennus villosus* just a few years ago. During our meeting I was also given the chance to talk to Iván Santibáñez Koref, who introduced me into the work and scientific findings regarding the *Cebrennus*. Furthermore Ulrich Berg presented me all the biomimetic robot implementations that were designed on the basis of the knowledge gained from the *Cebrennus villosus*. In this meeting with Rechenberg, I was finally able to validate all the information that I could only obtain from internet sources so far.

As I could find out from Rechenberg, he did not yet publish any documents or similar work to be used as a scientific source regarding his discovery on the *Cebrennus villosus*. Under this condition this book is going to contain the most recent research findings as well as being among the first scientific documents mentioning and building up on the work of Rechenberg, regarding the rolling locomotion of the *Cebrennus villosus*. The content and data presented within the chapters 2.1.1 through 2.1.4 of this book are therefore based on the interview with Rechenberg, except when other sources are disclosed. During our interview we also discussed some surrounding conditions and general assumptions regarding the spider. These and my own suggestions will round up the following chapters.

Figures that can be found in the following chapters were provided by Rechenberg himself in a PowerPoint presentation, which as well contains informative videos and high speed video captures of the *Cebrennus*. More figures were downloaded directly and extracted from more PowerPoint presentations available on the website www.bionik.tu-berlin.de, as recommended by Rechenberg. Further pictures shown in the following chapters were taken by me during the interview at TU Berlin.

2.1.1 *Behavioral Locomotion*

31° 16' 08,9" N; 03° 59' 28,7" W are the coordinates where Rechenberg annually puts up his camp during summer time. The place is located in the Erg Chebbi Desert in Southern Morocco, boundary to the Sahara Desert. It is here where he discovered and documented the rolling locomotion behavior of a desert spider in

Ralf Simon King: BiLBIQ: A Biologically Inspired Robot, BIOSYSROB 2, pp. 7–28.
DOI: 10.1007/ 978-3-642-34682-8_2   © Springer-Verlag Berlin Heidelberg 2013

2008. He realized far later that he had already documented the rolling locomotion of one specimen on video during his stay in 2006. Back then it was not obvious to him what the spider was really doing. He described the behavior for the first time in 2008 with new footage of the spider. While still wondered why it did not become obvious to him any sooner, he discovered a short rolling phase of the spider on an older film from 2006. This initial short-take was not recorded with a high speed camera. The first impression was that the locomotion of this spider is nothing special, because of the well-known Wheel spider. This spider from the Namib Desert in Southern Africa, also known as the Golden Wheel spider or Dancing White Lady spider, tumbles and rolls down desert dunes by curling up its feet and flipping to the side to escape predators. At first Rechenberg thought it might be a Wheel spider that got lost in the Moroccan desert.

Very soon it became clear to him that there is a significant difference between these two types of spiders. The spider investigated by Rechenberg suddenly started to roll from his hand to the plain desert ground after he released the specimen, he had caught the night before. To achieve this, the spider must be able to perform some sort of propulsion locomotion. This kind of locomotion performed by the spider was unknown within the nature's framework so far. Among other parameters, the most curious thing about this is that the spider performs this locomotion independently from surrounding conditions, which means that it does not need a slope to initiate the rolling process by using gravity force. Additionally the spider does not need to walk first or perform a startup gesture to trigger the rolling locomotion. It can roll instantly from the point where it stands. This became even more obvious when Rechenberg observed the spider rolling upwards a sand dune. The spider can perform this action on slopes of up to forty percent incline.[1] But with an increasing incline angle it can be observed that the spider slows down and becomes exhausted very quickly.

So far it could also be observed by Rechenberg that the spider starts to switch from a convenient walking locomotion to this unique rolling locomotion if very certain situations occur only. Such events might be the appearance of a predator, like a scorpion, or meeting a conspecific [if sex of the conspecific plays a role has not been researched so far]. The circumstance that the spider makes use of the rolling locomotion just to change position, hunt prey or to search for its tunnel, for instance, could not be observed either.

With these findings Rechenberg addressed the Berber people living in this part of the country. He explained and presented his observations to them by demonstrating a living spider performing its rolling locomotion. But this behavior seemed to be never observed by anyone even among the native people. An explanation for this could be that Rechenberg first became aware of this animal during a night walk through the desert. These spiders are active nocturnally or early in the morning before dawn to catch moths. During daytime they hide out in tunnels that they dug into the sand to escape the excruciating desert sun.

These tunnels which they need to dig out with their pedipalps [a set of tactile organs located on their heads] range twenty-five centimeters deep. Rechenberg also measured that they need to remove 80ml of sand for each tunnel structure of

---

[1] Rechenberg (2012) personal interview.

around two centimeters in diameter. Since the spider can only remove 0.1ml of sand with its pedipalps in a single excavation act, it needs to repeat this action eight hundred times to dig out a single tunnel. Moreover the tunnels are constantly being reinforced with the spiders silk already during the digging process. The fact that the tunnels are oriented vertically makes the reinforcement of the structure become even more significant. Otherwise the tunnel would collapse already after a few millimeters of digging due to the very fine, loose desert sand. After the spider enters the tunnel it closes the opening with a cap created of silk and sand. Within these tunnels the temperature stays constant at an approximate temperature of 35° Celsius during the day. This lets the spider's habitat become a quite well-tempered place compared to the top of the first layer of sand with temperatures of up to 70° Celsius during daytime. During his observations Rechenberg found that the spiders are able to find the ways back to their tunnels again, once they left them to go out hunting. This is very amazing because, as mentioned above, the spider covers its tunnel with the cap, thus making it a perfectly camouflaged spot and preventing sand from penetrating the tunnel. Since there are no visual signs due to monotony of the surrounding landscape, the spider's navigation capabilities become essential. Rechenberg suggests that the spider might use the stars for navigation, because it returns to the tunnel before dawn. During this time there is no sunlight available, as a result, no polarized light either, which is evidentially used by some insects and other animals for navigation. But up to now this statement has to remain with no scientific evidence. What can be clearly seen, however, is that the spider goes straight into the direction of its tunnel location and then, only on the last few meters or less, it starts searching fortuitous for it until it finds the tunnel cap.

Due to the circumstance that the intentional rolling locomotion of the spider was unknown until then, Rechenberg assumed it might be a subspecies that was not described so far. Accordingly he transported a living specimen back to Germany and gave it to the spider-expert Dr. Peter Jäger from the Senckenberg-Institute for closer investigation. Jäger was able to identify the spider as a huntsman spider of the genus *Cebrennus* species *villosus*. It was not the first *Cebrennus villosus* he had seen.

"From recent knowledge the distribution range of the genus Cebrennus reach[!] from Marocco[!] in the West to Israel in the East. In spite of their wide distribution range, Cebrennus spp are rarely found in museum collections. Reasons may be the cryptic way of living in silken retreats as well as the fact that they are nocturnal spiders."[2]

On a caught living specimen that was transported to Germany before[3], Jäger could not observe the behavior that Rechenberg described. In the interview with Jäger he confirmed that this behavior was not observed by him on the last specimen either and that, in addition to that, it was not observed by Uwe Moldrzyk, who thoroughly investigated the *Cebrennus* in Tunisia. Rechenberg and Jäger are

---

[2] Jäger (2000) p 163.
[3] cf Jäger (2000) p 184 f.

consistent in describing the villosus as a spider that is not easy to keep under artificial conditions. Rechenberg as well was rather unsuccessful in provoking the desired rolling behavior when keeping the spider in captivity back in Germany.

From a biological point of view the *Cebrennus villosus* still needs to be investigated more thoroughly. Should it be finally confirmed that the same villosus species, like the ones living in Tunisia, for instance, does not perform an intentional rolling locomotion, it might turn out, regarding to Rechenberg, that this species in Morocco is in fact a new subspecies. A conclusion like this is based on the assumption that it is hard to imagine finding the same body structure in the same spider species twice, even when living in most likely the same surrounding conditions, but still performing different locomotion behavior. A closer look might need to be taken on potentially different predators as well since the rolling behavior is considered as an escape strategy.

Due to the fact that this rolling locomotion must have evolved in an evolutionary process, based on the finding that not just one specimen but all offspring do perform this locomotion, it needs to be asked in what way the spiders from Morocco differ from the ones living in Tunisia, for example, in case the rolling behavior will not be observed there finally. The suggestion of considering the spider from Morocco as a new species, however, is not being traced further in this book and therefore considered to be a specimen of the *Cebrennus villosus*.

2.1.2 *Convenient Benefits*

If the spider uses a rolling locomotion to escape from a predator, for example, then there must be a specific benefit for those spiders using exactly this escape strategy. Otherwise evolution would have led this species to the path of extinction due to its bad performance and adaptation.

Rechenberg measured that in average the *Cebrennus villosus*, that is roughly the size of a human palm, can run with a speed of up to one meter per second. When switching the locomotion to rolling, speed doubles to two meters per second.[4] Hence **the spider's first benefit is – faster pace**. In another experiment he tested, which distance the spider can bridge using one technique or the other. The walking gait allows the spider to go as far as five to six meters. Then it needs to take a metabolic break of approximately three seconds before it is capable of running another five to six meters. When rolling, however, the spider can bridge a distance of up to twelve meters, before it has to stop for a break. Using the rolling locomotion allows the spider to make twice the distance within the same period of time. This makes it more efficient.

Assuming that the spider performs the locomotion of running and rolling until its exhaustion, the claim is derived that in case of rolling the spider is capable of bridging twice the distance within the same period of time yet using the same amount of energy. Hence **the spider's second benefit is – higher efficiency** through using the rolling locomotion. At this point it needs to be asked why the spider does not use the rolling mode all the time for position changing purposes.

---

[4] Rechenberg (2012) personal interview.

The answer might be that hunting or looking for mate, for example, is simply not possible when rolling. But these issues are to be solved through further research. At least the spider is more efficient in terms of time and distance - hence in speed.

2.1.3   Locomotion Analysis

By taking a look at the short-takes of the spider performing its rolling locomotion, apparently different types of motion can be observed. Abstract concepts describing the rolling could be: somersault; cartwheel; handspring; flic-flac. Unfortunately the description of what it is referred to by these words is incoherent. Therefore a quick description is made of what is finally meant when using these abstract concepts, referring to them being performed by a human. A somersault can be performed in forward or backward direction in the air or on the ground. A cartwheel is performed in sideward direction with ground contact of each limb after another. The terms handspring and flic-flac are considered to be on the same level, meaning a flip to the front or back while hands simultaneously touch the ground, then the body flips over simultaneously landing on the feet again. While Rechenberg is talking mostly about a "Salto", which is the German expression for a somersault in the air, Jäger prefers to describe it as a "Flick-Flack" meaning a flic-flac or handspring in forward direction.

Fig. 2.1[5] *Cebrennus villosus* in different states of the rolling locomotion. Figures are stitched together and do not show the same movement performed at one time.

A cartwheel can obviously be excluded from the analysis because the spider does not flip to the side. A somersault on the ground can be excluded as well due to the reason that the main body of the spider has no contact with the ground like a human that would roll over his back. The comparison of a human and a spider performing this motion is not easy, though. A human that performs a somersault in the air starts to jump using the bounce of his legs and then lands on his feet again similar to the spider, whose body faces to the ground when triggering the jump and doing most likely the same upon landing. The observations made by Rechenberg – namely that the spider performs a flip of almost 360° most of the time - promote the idea of a somersault. According to him, the spider only uses the last pair of legs when it becomes exhausted, while the first and second pairs of

---

[5] cf figures Rechenberg (2012) PowerPoint and `http://www.bionik.tu-berlin.de`. Accessed 06 Jan 2012.

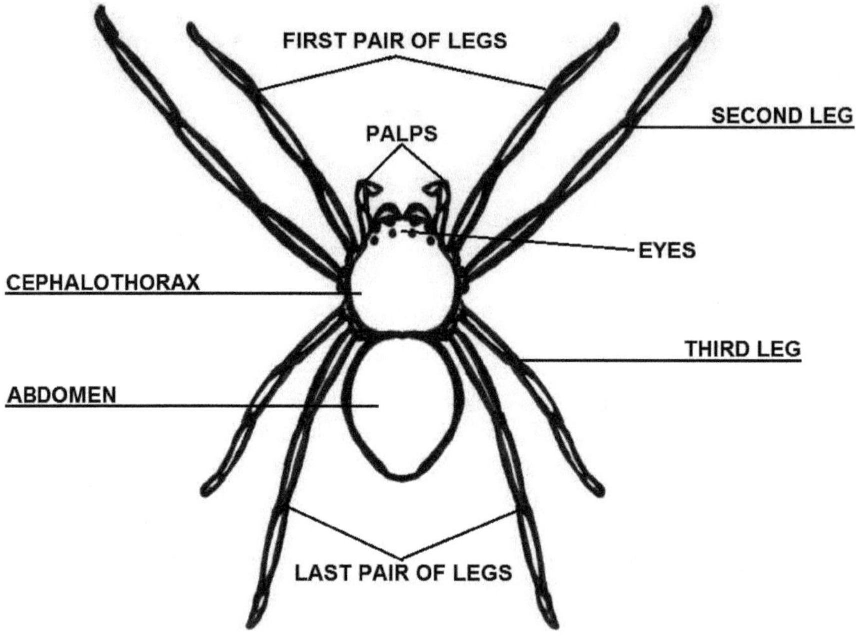

Fig. 2.2 Figure of a spider displaying where the leg pairs are located

legs are obviously involved in the propulsion process. In extreme cases, when movement becomes very clumsy due to exhaustion, even the third pair of legs would touch the ground.

However, it could be hold against this argument that the spider lands on a different pair of legs first, compared to the pair of legs which had the last ground contact when jumping. But to prove this, a closer look on the movement of the spider needs to be taken. If the assumption turns out to be true, this would be a strong argument for a handspring. The fact that the spider first needs to flip over to the front before it can trigger the jump would be another reason to call the movement a handspring. The spider also seems to use the kinetic energy gained from its last move to partially load for the next one. That the spider is capable of preloading its hydraulic system [see chapter 3.2.4], or storing energy in its muscles when landing, can merely be suggested - but not proven. The usage of the momentum of the last flip seems to be reasonable but providing evidence for this is not in the scope of this book.

When a human performs a somersault in the air, he can either perform it once and needs to stop then, or, with support of thirds, perform multiple, consecutive, somersaults in the air without touching the ground once. But once the person lands on their feet, the energy is dampened because humans are not predestined for moves like preloaded jumps. This becomes different when performing a handspring. In a handspring the stored energy in muscle fibers cannot be used, for instance, but the momentum of the handspring performed before can be used and converted into an even faster movement upon the next turn. This given fact would foster the tendency to call the motion performed by the spider a handspring. The comparison and evaluation of these terms and definitions unveils the complexity of the locomotion itself. As for now it cannot be discussed in every detail, but a final decision up on the term needs to be made. Since the spider has only legs and no arms and hence no hands, it appears to be contradictory to call the locomotion a handspring. Therefore the term somersault, meaning a somersault in the air, is being used from now on.

Rechenberg was able to record sequences of the spider doing somersaults. Unfortunately the footage is very blurry. Rechenberg had to use a low cost compact camera he could set to a maximum speed of 480 frames per second with the tradeoff of a reduced resolution. The single frames shown in his PowerPoint

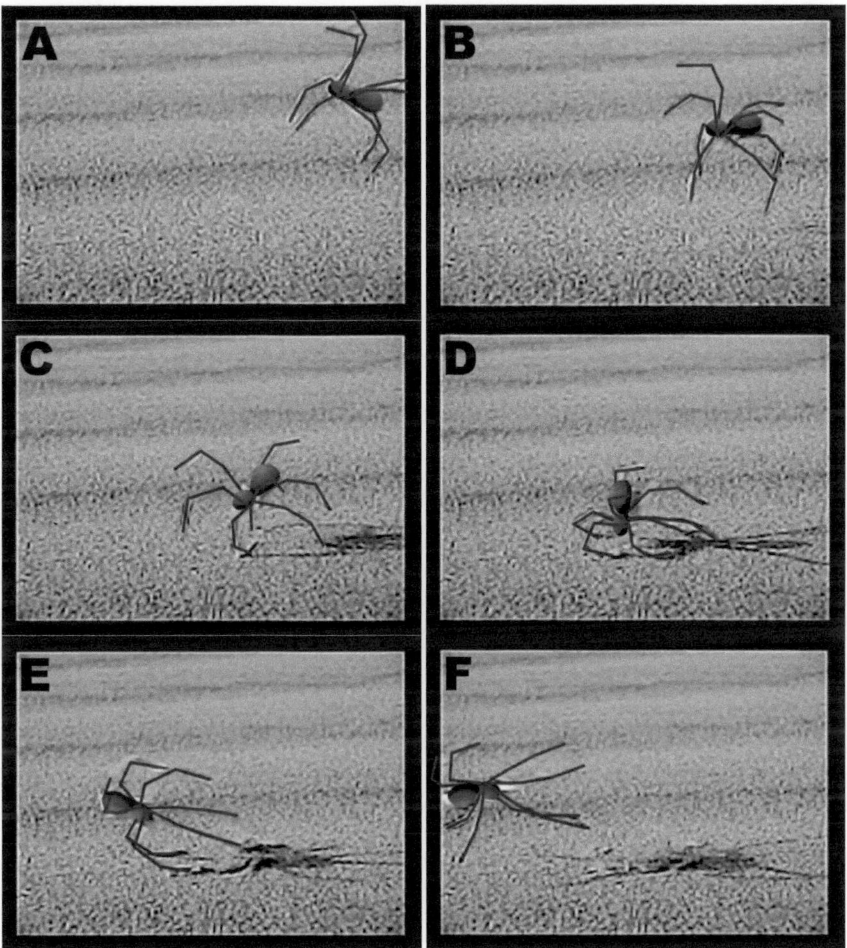

Fig. 2.3[6] Figures A to F taken from the somersault sequence. Spider is already in rolling mode, lands on its legs and triggers a new jump.

---

[6] cf figures Rechenberg (2012) PowerPoint.

presentation were overlaid by him with outlines so that the movement and position of the spider's body and limbs could be distinguished from the background more easily. Even though the limbs can be seen better in the figures now, it is still hard to track which pair of legs plays which functional role in this special movement.

The implementation of the spider movement into a robot model requires a thorough look on what role the individual pairs of legs play in the locomotion. The knowledge gained by precisely observing the movement might turn out beneficially when building the robots structure later. Since position tracking of the single pairs of legs is still hard, I overlaid the limbs with different colors in the following figure.

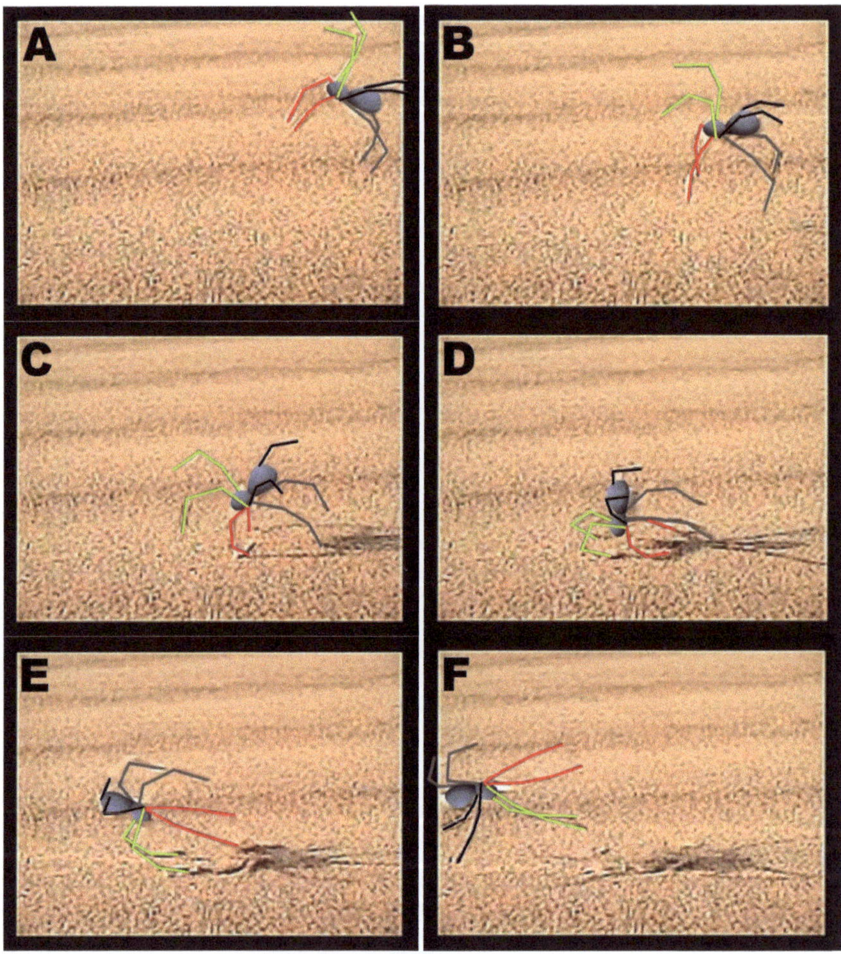

Fig. 2.4[7] Sequence overlaid with colors to distinguish better which pair of legs does what

---

[7] cf figures Rechenberg (2012) PowerPoint.

Scene (A) of the figure above shows the spider being in the air with the first pair of legs [red] pointing downwards. The second pair [light green] and third pair of legs [black] point upwards and the last pair of legs again points downwards. In scene (C) the spider seems to have initial ground contact with the first pair of legs followed by the second pair of legs in part (D) when landing. Finally the first pair seems to give a high propulsion impulse to the spider and the second pair of legs is the one that has the last ground contact. According to Rechenberg this cannot be true. He stated that the second pair of legs is the one playing the essential role in the propulsion and I observed that the first pair of legs always seems to be the pair with the last ground contact. I therefore consider these with outlines overlaid images of the PowerPoint sequence as slightly imprecise, but it might also be the low video quality that leads to a misconception here. In two individual pictures with a clear shot of the movement it can be clearly claimed which pair of legs takes which position. Again I highlighted the legs shown in the figure below. I also assumed that the spider performs a somersault to the front in these cases.

Fig. 2.5[8] Spider in motion, first pair of legs in red, second in light green, third in black and fourth in white

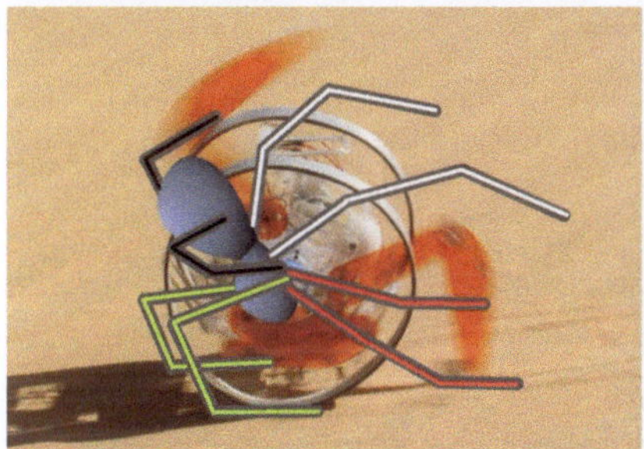

Fig. 2.6[9] The first pair of legs (red) should not point downwards as the left figure implies (background shows a robot implementation). The right figure shows the first pair of legs (red) in an upper position relative to the spider's body and the second pair of legs (light green) with a high propulsion impact on the sand.

---

[8] cf figures Rechenberg (2012) `http://www.bionik.tu-berlin.de`. Accessed 06 Jan 2012.

[9] cf figures Rechenberg (2012) PowerPoint and `http://www.bionik.tu-berlin.de`. Accessed 06 Jan 2012.

The misconception can clearly be observed using a direct comparison of an overlaid image of the PowerPoint presentation by Rechenberg, to which I add colors, in turn, contrasting a clear shot of the spider.

By further observation of the PowerPoint sequences downloaded from www.bionnik.tu-berlin.de and the short-takes provided by Rechenberg, some interesting aspects could be found. Among those the spider curls up during its flight phase. To what extent this might influence the locomotion performance and in which way the spider makes use of a preloaded system, or if it is able to use energy from landing for the next jump can only be estimated for now. An interesting aspect still remains the one whether the spider prepares for the rolling phase by positioning its first pair of legs to allow a rollover. The following figure shows a composition of screenshots extracted from one of the slow-motion short-takes.

This might be a first hint to a good robot implementation. Due to the reason that the spider can trigger the rolling locomotion without the necessity to run first the decision is reasonable to realize this movement in a non-dynamical and therefore well controllable way. In scene (B) of Fig 2.7 the last state of the walking locomotion can be seen. In scene (C) the spider prepares for a jump over the front by positioning the first pair of legs [red] right in front of it ramming them into the sand. The second pair of legs [light green] bends and is located to the left and the right of the spider. In scene (D) of the figure, the second pair of legs stretches and lifts the fourth pair of legs [white] from the ground, while the third pair of legs [black] has already lost ground contact and curls up closely to the body of the spider. In scene (E) the fourth pair of legs curls up closely to the body as well, while the second pair of legs is almost fully stretched. The first pair of legs stretches as well and is the last one having contact to the ground. Fully airborne in the last part of the figure, the spider has already performed a 180° flip and now faces the camera. 180° later it lands on the second pair of legs first, then on the first pair and triggers another jump. The noticeable thing about this is that until the very last moment, where the spider starts to be more airborne than having ground contact, it still has four legs on the ground. This is beneficial for implementing a stable motion with a robot model like the one mentioned earlier.

Other aspects remain hard to implement such as performing the following flight phase or make use of a mass-spring like system. Therefore I will not go deeper into the analysis of the movement. I will also disregard the walking gait of the spider, because I neither have specific information about it nor a specimen to examine it on. This would be necessary to find out more about potential special movement patterns due to the specific leg configuration of the spider.

The complexity of the whole system becomes more obvious by thinking about the anatomy of the spider. Measuring contact forces on the surface or how much pressure the spider applies on the different pairs of legs with its hydraulic system would also be an insightful research topic. Internal liquid pressure is of course very hard to measure, especially because the spider still needs to be able to

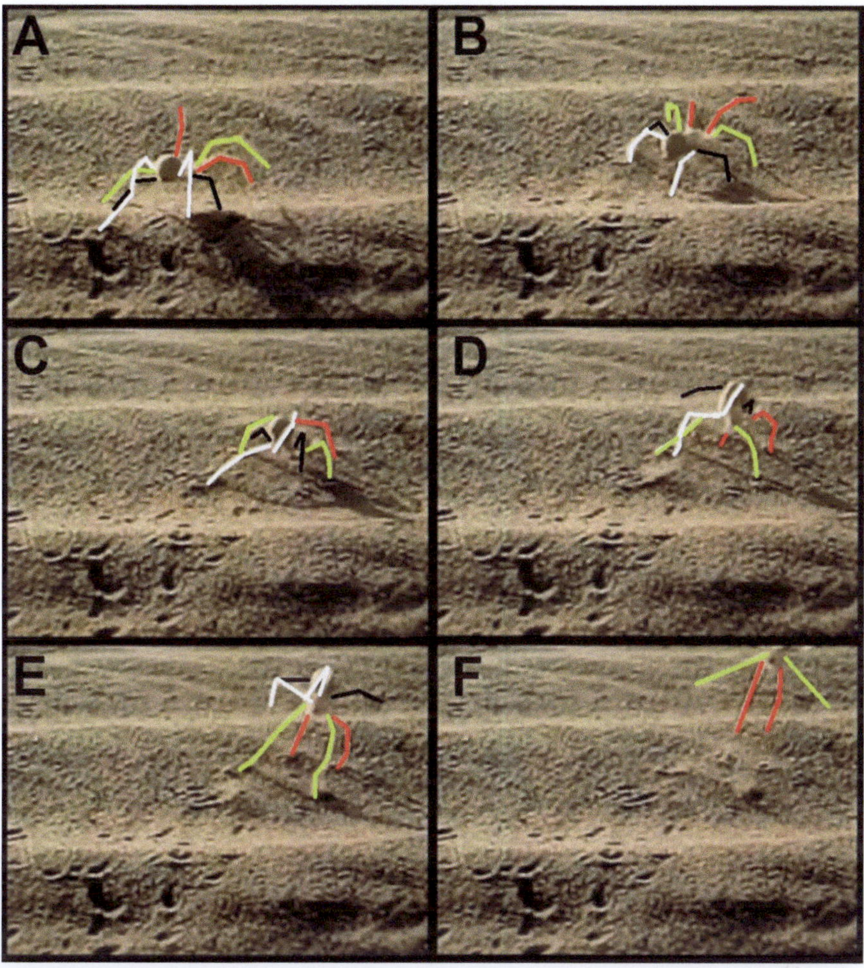

Fig. 2.7[10] Screenshots from a short-take showing a *Cebrennus villosus* switching its locomotion. First pair of legs in red, second in light green, third in black, fourth in white.

perform its movements undisturbed. Depending on the anatomy, for example how large the flexor and extensor muscles are, more understanding will be gained about the role the single limbs play. As for the moment the difference in the length of the legs can provide some more information:

Regarding to Rechenberg it is interesting that the *Cebrennus villosus* has a different set of legs compared to a typical jumping spider. Even though the *Cebrennus villosus* does perform jumps during the rolling locomotion, it is the second pair of legs that is the longest, compared to the jumping spider with its fourth pair of legs as the longest. Among spider-experts the length of the leg pairs

---

[10] cf short-take Rechenberg (2012) PowerPoint.

is one feature that allows classifying spider types. The length of the legs is encoded in a string of four digits. For the *Cebrennus villosus* this would be: $(2413)^{11}$. The positions of the digits represent the position of the leg pair. Hence the first number represents the first pair of legs. The value of the number indicates the relative length of the leg pair. Hence a (1) stands for the shortest pair of legs and a (4) stands for the longest pair of legs of the specimen. This might be evidence for considering the second pair of legs, which is the longest, as the one producing the most propulsion when performing the rolling locomotion.

In consideration of how relevant the limbs' positions are for an advanced dynamic robot implementation, I suggest an improvement of the method the spider is observed with. The usage of a professional high-speed camera with an adequate resolution would be an easy but cost-intensive improvement. Instead, or additionally, the examined spider specimens could be painted or prepared with markers on the joints and tips of their legs to help distinguish their positions when moving more easily.

After the analysis considering the documents and thoughts Rechenberg provided, another reason is found to call the locomotion a somersault: The spider is capable of performing a full rotation in which the third and fourth pair of legs is mostly not in use. It is assumed that the spider uses the fourth pair of legs every now and then to control and/or correct the direction of its rolling motion. The fact that the spider is capable of setting and adjusting a desired curve when rolling is undoubted due to the observations. A very basic but important finding is that the spider is forming an abstract wheel with its legs, a wheel that is neither mounted on, nor spinning around a fixed axle. Hence the whole body structure is rotating during the rolling locomotion.

Fig. 2.8[12] *Cebrennus villosus* positioning its legs and forming an abstract wheel

---

[11] cf Jäger (2000) p 183.

[12] cf figure Rechenberg (2012) PowerPoint and `http://www.bionik.tu-berlin.de`. Accessed 06 Jan 2012.

2.1.4 Robot Implementation

Rechenberg already implemented the knowledge he gained into a set of robots. All
the robots obey the fact that the spider is not spinning around a fixed axle and that
the whole body is performing a rolling motion. Since the spider has eight legs that
it can bring into the shape of a wheel it needs to be asked which form could mimic
that. Starting with a simple single actuator robot [I. in the following figure] the
models transformed more and more towards the biological sample.

Fig. 2.9 Set of robots developed by Rechenberg and his team

The first robot (I.) has a cylindrical shape with a single actuator. Once
the rotation is initiated manually, a momentary contact switch will trigger an
actuator moving out of the cylinder and forcing it into a continuous rotation.

The momentum of the spin is high enough to trigger the switch again and the cylinder keeps rotating straight forward. The second robot (II.) works similarly but comprises two actuators and two momentary contact switches, which are mounted respectively opposing. The cylindrical form is replaced by an oval shape, which ensures that the robot will flip on the long side of the oval through gravity force where the switch will trigger the actuator to drive it. The robot instantly starts to roll once being disengaged due to its oval shape. Even though the third generation of the robot (III.) has just one additional actuator and one momentary contact switch each, it basically follows a more complex idea.

Iván Santibáñez Koref, from TU Berlin, explained to me that the idea is like drilling a rectangle. At first glance this seems to be contradictory, but with a driller shaped like a Reuleaux-triangle a satisfyingly rectangular shape can be achieved. The secret behind the Reuleaux-triangle is that it maintains the same height like a wheel when rolling, but still has an irregular shape. Another special feature about the wheel is that its center does not stay in the middle of two parallels thus preventing the use of a Reuleaux-triangle on a fixed axle on the one side but making it a perfect rolling device on the other.

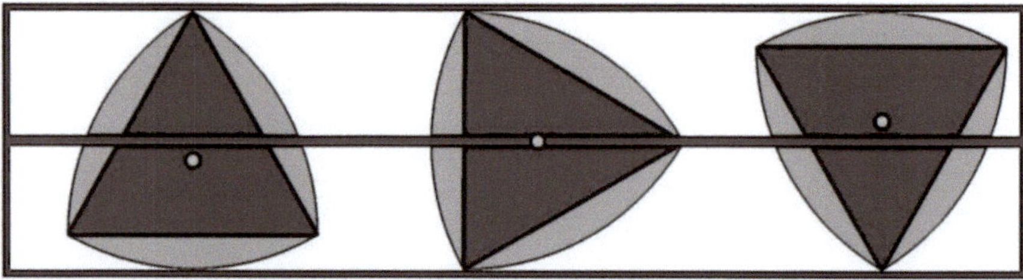

Fig. 2.10 CAD visualization of a Reuleaux-triangle in different positions. The center of the triangle is clearly moving eccentrically while the Reuleaux shape maintains the same height.

The left robot in the following figure was tested successfully in the Erg Chebbi Desert. The Reuleaux-triangle's shape ensures that the robot always flips onto one of the momentary contact switches, thus triggering the actuator to drive the robot beyond the death point of the rolling movement and hence onto the next switch.

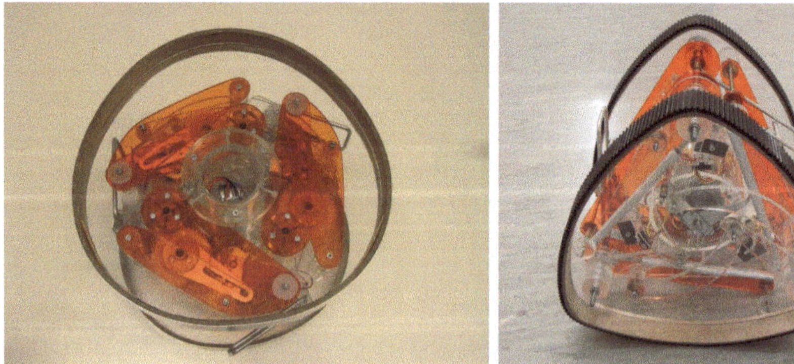

Fig. 2.11 Two similar robot implementations of the Reuleaux-triangle

The latest robot [IV. in Fig 2.9] has six legs in three pairs, which takes it further towards the spider with its four pairs of legs. This robot is radio-controlled and the possibility to control the direction of its rolling to the left and right is suggested. But a first implementation did not work as expected. The capability of rolling backwards is not given yet. The latest robot mimics the spider movement in a way that some torque applied on the pair of legs facing the ground, leads to an instant rollover. This also enhanced the speed of the robot.

The rolling locomotion is fulfilled with Rechenberg's robots, even though they are not performing a somersault in the air like the spider does. The full rotation somersault on the ground with just triggering an actuator once is also not given in the latest models - but in the first, where a cylindrical shape was used. One aspect, however, was completely disregarded up to present: The spider can also walk in fact. The robots shown above are not capable of performing a walking gait. Their structure is completely designed to fulfill rolling locomotion. Hence the focus here is to realize a rolling locomotion similar to the one the spider performs. Therefore, the development of a satisfying rolling and stable, non-dynamical walking locomotion is the aim of this book.

2.1.5 A Technical Pendant

An interesting approach that shows some parallels with the basic abstract principles of the spider locomotion is the 'Aktivspeichenrad' or ASR as the inventor calls it. 'Aktivspeichenrad' is German and would, in a literal translation, mean 'active spoke wheel'. This invention by Michael Post is roughly described as a wheel that is driven directly through its active spokes using linear actuators. Post already invented the wheel and applied for a patent in the year 1999, when the spider locomotion was not discovered yet. His patent was finally granted by the German patent office in 2009.[13] Referring to the interview with him, the patent office even had to introduce a new category for his invention. One feature of this invention is that the center of the wheel does not rest in the middle of the object to perform a rolling locomotion, similar to the center of the Reuleaux-triangle. In the Reuleaux-triangle the center is fixed and moves eccentrically due to its external shape. On the active spoke wheel the outer shape is a circle but the center moves actively with the help of its linear actuators. A significant difference to the biomimetic implementation is that the active spoke wheel needs to be mounted on an axle that allows it to fulfill the eccentric movements. Hence, practically two active spoke wheels are needed to perform a rolling locomotion with one of them being fixed to a platform and the other one acting as the actual wheel. An advantage of this approach is that this type of wheel can be used on any kind of vehicle in contrast to the Reuleaux-triangle solution.

---

[13] cf Post (2009) patent.

Fig. 2.12[14] Active spoke wheel on a futuristic car

It could be suggested that a single active spoke wheel, for example with three linear actuators, might be used without an axle. The structure to which the axle would usually be fixed has a certain weight due to carrying a battery, for instance, which then would be heavy enough to generate a momentum to the wheel when moved out of the center of gravity. The linear actuators would keep the center of gravity outside the geometrical center of the wheel and therefore lead to a rolling locomotion. This motion would result in a rotation where the whole structure would circle around an eccentric axle, similar to the Reuleaux-triangle. However, this suggestion applies only to a robot implementation that can move freely and independent from a platform. Using it for transportation purposes, especially when mounted to a vehicle, becomes impractical then.

2.2 Further Rolling and Tumbling Locomotion

The following examples that were found during the research phase should only give a brief look at further biological samples. They represent the most interesting solutions of nature all sharing a certain kind of rolling or tumbling locomotion. The *Cebrennus villosus*, however, still holds a unique and unreached position thanks to its ability to walk and perform an intelligent and directed rolling locomotion. Information taken from books that are based on internet sources as well as suitable internet sources comes to use here.

---

[14] Figure provided by Michael Post via email.

2.2.1 The Wheel Spider 'Carparachne aureoflava'

As mentioned earlier, the very first impression Rechenberg got from the spider
was that the *Cebrennus villosus* simply performs a motion similar to the well-
known Wheel spider. The Wheel spider can be found in the Namib Desert of
Southern Africa. Like the *Cebrennus villosus* it is a nocturnal spider living in
silken retreats. If the Wheel spider is threatened by a predator that it cannot fight
off, and the spider is positioned on a sloped dune, it will flip to the side and roll
down the dune using the force of gravity performing cartwheels.[15] So far no
further hint was found, neither in the scientific literature that was available to me,
nor during my interviews, promoting the idea of the Wheel spider using a self-
propulsion technique for its rolling locomotion.

2.2.2 Comparison of 'Cebrennus villosus' and 'Carparachne aureoflava'

After the locomotion analysis of the *Cebrennus villosus* and the comparison
between the two spider types it needs to be mentioned that the robot
implementation of the project before matched rather the locomotion of the
Carparachne aureoflava than the locomotion of the *Cebrennus villosus*. Even
though the robot had an own propulsion system it was still performing a
cartwheel. The approach can be comprehended as described in the following:
    While the *Cebrennus villosus* is performing a highly dynamic locomotion, a
robotic implementation with a well controllable, stable behavior was sought for.
With no robotics- related background and experience, in terms of non-industrial

Table 2.1[16] Comparison of the two rolling spider types

Spider Genus	*Cebrennus villosus*	*Carparachne aureoflava*
Type	Huntsman	Huntsman
Active time	Nocturnal	Nocturnal
Retreat	Silk-lined burrow, vertical	Silk-lined burrow, angular
Spider size	100 mm	20 mm
Running speed	1 m/s	Not specified
Rolling speed	2 m/s on plain ground	1 m/s downhill
Locomotion	Somersault in the air / Handspring	Cartwheel

---

[15] cf Hephaestus Books (2011) p 42.
[16] Rechenberg (2012) personal interview; cf Hephaestus Books (2011) p 42.

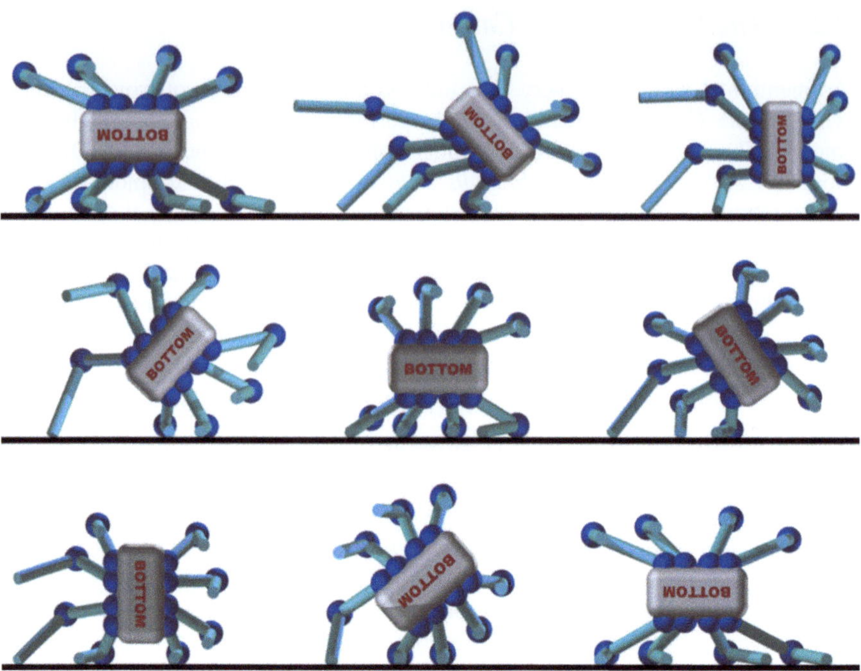

Fig. 2.13 CAD visualization of a robot that performs a 360° turn on its side

robots, a first approach was to let the robot kneel to its side when preparing for a rolling locomotion. Rolling would then be achieved by bending and stretching the accordant legs.

The figure above displays a robot configuration in which the robot has eight legs accordingly to the spider. For the CAD robot visualization the smallest possible amount of joints per leg was suggested with two for a reduction of complexity. Each joint, resembled by a blue ball, has three rotational degrees of freedom [DOF]. The resulting six DOF per leg cannot be compared directly to those of the spider because: The exact amount of DOF that the *Cebrennus villosus* has was not found out during the research work and is thus an unknown factor. Due to the complexity that arises from the fact of having eight legs a structure was designed to simplify the robot by reducing the number of legs to four. In contrast to Fig 2.13 the necessary DOF per leg were assessed with four. Each joint has now one rotational DOF. Three are necessary to achieve a linear movement of the last link of the leg, while the fourth DOF, would allow the robot putting its leg forward or backward during a walking gait.

This robot was expected to be able to switch from the walking gait to the rolling gait and back without external aid. Besides the capability of the robot to roll in a forward and backward direction, the direction of rolling could have been controlled due to its four DOF per leg. This is because if the last link can be

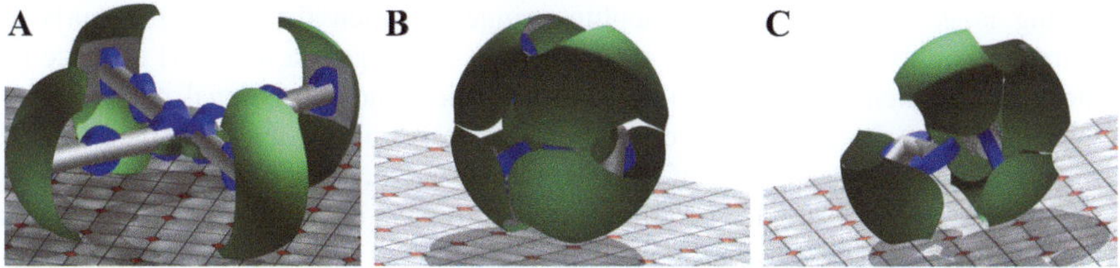

Fig. 2.14 Schematic CAD visualization of the rolling robot in configuration (A) walking position, (B) rolling position and (C) rolling by pushing one leg against the ground to achieve propulsion

moved in a linear dimension, then it can also be moved in a certain angle and therefore steer the robots direction already during the propulsion movement. As explained in the introduction this robot failed in performing a proper walking gait due to an ineffective structure design and to hardware constraints. Nevertheless the rolling locomotion worked in principle and could have been improved with a position-sensor and accordant programming.

2.2.3 The Mount Lyell Salamander

Another animal performing a passive rolling locomotion is the salamander *Hydromantes platycephalus* observed in the northern Sierra Nevada in California. This salamander is curling up its body, tail and legs as a defense mechanism. It was recorded that the salamander remains in this spherical shape for three to ten seconds. In case of placing the salamander on a slope, the spherical shape results in a rolling movement of the salamander. This is different compared to other salamanders. While curling up as a defense mechanism is quite common among other species of salamanders, it was observed that these in contrast prevent rolling by not completely curling up their legs or tail, for instance. Curling up the body entirely is not wide spread among animals and only a handful of species actually show the rolling behavior. Based upon research it can be claimed that all curled up specimens of the *Hydromantes platycephalus* would start to roll when being positioned on a slope. In addition to that they remain in the spherical shape until their body comes to a complete stop when the slope ends. Then, after remaining in the normal state for approximately five seconds, the salamander starts seeking shelter in a normal running manner.[17]

2.2.4 The Pearl Moth Caterpillar

The caterpillar *Pleuroptya ruralis* performs a backward roll when poked. It curls up into a ball and revolves for approximately five times.[18] The energy needed for

---

[17] cf Garcia-Paris, Deban (1995) p 149 f.

[18] URL: http://www.abc.net.au/science/articles/1999/08/09/ 42510.htm?site=science/greatmomentsinscience. Accessed 05 Feb 2012.

rolling is gained from the momentum triggered when the caterpillar curls up extreme quickly. The caterpillar needs to relax and trigger a new curl up once the momentum taken from the curl-up is transformed into a rotational movement of the caterpillar. It does not need a slope to start rolling and can therefore be regarded as one specimen of the group of animals with an active rolling locomotion. This sample of nature has already been implemented into a biomimetic artificial structure.

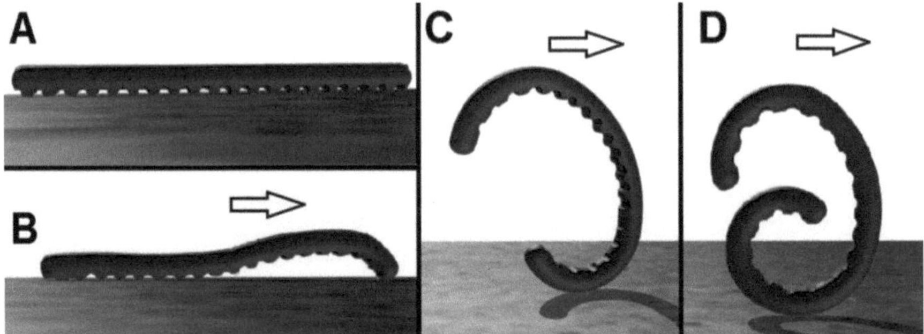

Fig. 2.15[19] Schematic visualization of the GoQBot soft-bodied robot, simulating ballistic rolling behavior. Black arrow implies direction of movement.

However, not only the *Pleuroptya ruralis* but other caterpillars as well are capable of rolling away from a predator when threatened. The biomimetic robot in the figure above was realized with SMA coils [see chapter 3.2.2] as actuators and silicone to form the body. For convenience reasons the robot performs a motion in forward direction only, in contrast to the caterpillar, which rolls in backward direction. The robot has no righting reflex to turn back in its initial position independently so far. The GoQBot is able to completely curl up within 200 ms, which is as half as fast compared to what the caterpillar under research here can achieve. The GoQBot then reaches a velocity of more than 0.5 m/s. This can only be compared to the caterpillar with respect of size, since the animal can achieve a velocity of more than 0.2 m/s with a body that is roughly five times smaller than that of the robot implementation.[20]

2.2.5 The Mantis Shrimp

Another actively rolling animal is the *Nannosquilla decemspinosa*. The leg structure of this marine animal is not suited for locomotion on land. Hence it performs a series of backward somersaults once washed to the beach, seeking to go back into the sea. The tested specimens with a maximum length of twenty-three millimeters can perform twenty to forty somersaults in a row. In average the

---

[19] After figure 1, Lin, Leisk, Trimmer (2011) p 2.
[20] cf Lin, Leisk, Trimmer (2011) p 2 ff.

shrimp can bridge a distance of one meter, but a maximum continuous trip of more than two meters was recorded. A maximum speed of 0.056 m/s was recorded during the rolling locomotion. The shrimp is also able to overcome an uphill slope of 10° on the beach and of up to 30° incline under laboratory conditions. It could not be proven, however, that the Mantis shrimp guides the direction when it starts to roll. There is just a small tendency that it prefers to go downhill on the beach – a behavior that in fact would lead it back to water more quickly.[21]

2.2.6 A Rolling Plant

The tumbleweed is a plant that is well-known from western movies. Once it is full-grown it dries out and disengages from its root in the soil. The dead tumbleweed will then, driven by the wind, roll over the land and rather loose than actively spread its seeds.

NASA´s JPL has developed a robot that obeys the main principle of the tumbleweed. This lightweight robot would be driven by the wind as well and offer some advantages compared to other rovers that were send to space before. The tested robot reached speeds of up to thirty kilometers per hour when there was wind. This is quite fast compared to the Mars rovers Spirit and Opportunity, which can reach circa 0.0139 m/s in average.[22] Other institutions as well work on tumbleweed-like robots, but initially the 'Tumbleweed rover' of NASA's JPL was not a brainchild with the biological tumbleweed as a sample.

"The Tumbleweed currently under development at JPL has a rather serendipitous origin. While researchers in JPL's Inflatable Technology for Robotics Program were testing a three-wheeled, inflatable rover in the Mojave Desert, one of its 1.5 m diameter nylon tires was detached by a gust of wind. The runaway tire quickly picked up speed in the moderate wind and seemed relatively unimpeded by the desert's rough terrain. The renegade ball was able to climb steep slopes, over large boulders, and through the jagged brush without hesitation. This seemingly unlucky incident produced a rather lucky discovery and was the inspiration for the current Tumbleweed rover."[23]

2.2.7 Résumé of Nature's Archetypes

Further examples of rolling animals like the pangolin[24], or other biological forms, may it be with the help of gravity, with an own propulsion method or other approaches, were considered as not relevant for this book. An additional reason for doing so is given by the description of an animal performing a rolling locomotion, for escape purposes, for example, appeared to be happening just

---

[21] cf Caldwell (1979) p 71 f.

[22] URL: `http://astrobio.net/pressrelease/861/tumbleweed-rover`. Accessed 12 Dec 2011.

[23] Behar, Matthews, Carsey, Jones (2004) p 4.

[24] cf Hephaestus Books (2011) p 33; Surhone, Timpledon, Marseken (2010) p 2.

randomly or descriptions relied on statements made by individual persons only. Moreover, the contribution to this book was considered relatively low, compared to them gained from the advanced abilities of what the spider *Cebrennus villosus* can do.

References

Behar, A.E., Matthews, J., Carsey, F., Jones, J.: NASA/JPL Tumbleweed Polar Rover (2004), `http://hdl.handle.net/2014/39171`, `http://trs-new.jpl.nasa.gov/dspace/bitstream/2014/39171/1/04-0014.pdf` (accessed February 12, 2012)

Caldwell, R.L.: A unique form of locomotion in a stomatopod – backward somersaulting. Nature 282, 71–73 (1979), `http://www2.nau.edu/~bio222-c/Reserve%20Reading/RR6/Caldwell_1979.pdf`, doi: 10.1038/282071a0 (accessed February 12, 2012)

Garcia-Paris, M., Deban, S.M.: A Novel Antipredator Mechanism in Salamanders: Rolling Escape in Hydromantes platycephalus. Journal of Herpetology 29(1), 149–151 (1995), `http://debanlab.org/storage/GarciaParisDeban1995.pdf` (accessed February 12, 2012)

Hephaestus Books: Rolling animals, including: Armadillo, Hedgehog, Pangolin, Armadillo Lizard, Rough-skinned Newt, Chinhai Spiny Newt, Mount Lyell Salamander, Pleuroptya Ruralis, Rotating locomotion in living systems, Wheel Spider. Hephaestus Books, USA (2011) ISBN-13:978-1243455659

Jäger, P.: Phone interview (January 06, 2012)

Jäger, P.: The huntsman spider genus Cebrennus: four new species and a preliminary key to known species (Araneae, Sparassidae, Sparassinae). Revue Arachnologique 13(12), 163–186 (2000), `http://www.senckenberg.de/root/index.php?page_id=144&standort=true&sektionID=50&abteilungID=1&standortID=1&showPageID=528`, `http://www.senckenberg.de/files/content/forschung/abteilung/terrzool/arachnologie/cebrennus_jaeger_2000.pdf` (accessed February 09, 2012)

Lin, H.-T., Leisk, G.G., Trimmer, B.: GoQBot: a caterpillar-inspired soft-bodied rolling robot. Bioinspiration & Biomimetics 6(2) (2011), `http://iopscience.iop.org/1748-3190/6/2/026007`, doi:10.1088/1748-3182/6/2/026007 (accessed February 05, 2012)

Post, M.: Phone interview (January 09, 2012)

Post, M.: Aktivspeichenrad. German patent DE000019957373B4 (July 23, 2009)

Rechenberg, I.: Personal interview (January 27, 2012)

Surhone, M.L., Timpledon, M.T., Marseken, F.S. (eds.): Rotating Locomotion in Living Systems: Rotation, Rolling, Wheel, Animal, Locomotion, Natural Selection, Evolution of Flagella, Robot Locomotion, Suspension (Vehicle). Betascript Publishing, USA (2010) ISBN:978-613-0-31606-8

`http://www.bionik.tu-berlin.de` (accessed January 06, 2012)

`http://www.abc.net.au/science/articles/1999/08/09/42510.htm?site=science/greatmomentsinscience` (accessed February 05, 2012)

`http://astrobio.net/pressrelease/861/tumbleweed-rover` (accessed December 12, 2011)

3 State of the Art in Robotics and Robotic Actuation

3.1 Existing Walking and Rolling Robots

The variety of robots that are able to walk or to roll is enormous. There is an immense number of books dealing with the problem of walking gaits alone, may the robot have one or eight legs. Literature on designing rolling robots is also available in an ample variety. Due to this matter this chapter focuses on the most interesting results of the research work. Among these there are mostly robots combining different locomotion, or performing a gait in a unique way.

Robots that have wheels mounted to their legs are categorically excluded from the research because this would mean having an additional structure, which is contradictory to the aim of performing two very different kinds of locomotion with one and same structure. So called hybrid legs [legs that revolve continuously because they are mounted to an electrical motor, for instance] or hybrid wheels [wheels that can open up to form a leg] count to the category disregarded as well, even though there are both some quite interesting approaches towards this and good locomotion performances were achieved with prototypes. Passive wheels used for a skating motion performed by a multi-legged robot are also not considered here. In summary, whatever kind of wheel we might deal with, it would still be one that moves around a fixed axle. It is determined earlier that this is contrary to the biological archetype and was yet not found within the framework of nature's inventions. Some of the following robot examples are not taken from primary sources. However, this shall not be considered an issue because I only want to give an impression of the current state in robotics. The question thereby is, whether there are already existing robots that can perform walking and rolling, with the same structure and alignment of actuators, in an effective way.

3.1.1 *Jumping and Crawling Robot*

One approach to combine different ways of locomotion is sought to be solved with the usage of SMA coil actuators as displayed in the figure before. Arranged like spokes in a spherical or wheel-like soft bodied structure, they can change the profile of the outer hull and therefore lead to a movement of the structure. "The circular prototype moves about 65% of its diameter per second, climbs up a slope of 20°, and jump[!] twice its diameter."[1] However, these prototypes are still in a

---

[1] Sugiyama, Shiotsu, Yamanaka, Hirai (2005) p 3611.

Ralf Simon King: BiLBIQ: A Biologically Inspired Robot, BIOSYSROB 2, pp. 29–47.
DOI: 10.1007/ 978-3-642-34682-8_3   © Springer-Verlag Berlin Heidelberg 2013

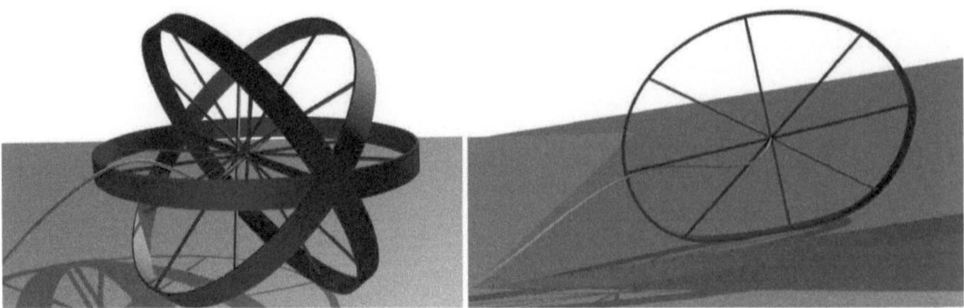

Fig. 3.1[2] *(left)* Spherical prototype for crawling and jumping. *(right)* Circular robot climbing a slope.

very early stage of development. For the time being they are not able to carry a controller and a power source to drive the SMA coils. Therefore they cannot move independently and outside laboratory conditions.

3.1.2 *Jumping and Rolling Robot*

A similar yet different idea is thought by Rhodri Huw Armour. He designed a spherical robot called "Jollbot" that can roll in any user desired direction and additionally jump to overcome obstacles for instance. Jumping trajectory can be guided into any desired direction as well. Furthermore the robot is capable of rolling up slopes of four degrees[3] having its center of gravity out of the center of geometry. His latest version, named Jollbot 3b, is also able to carry a payload of

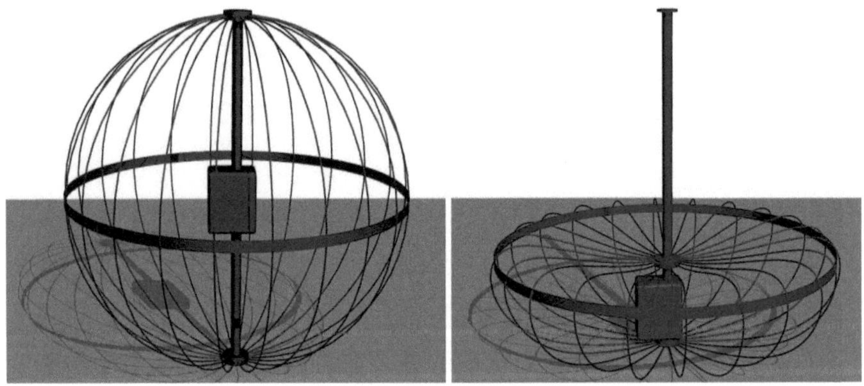

Fig. 3.2[4] *(left)* Jollbot 3b with actuator centered. *(right)* Jollbot ready to trigger a jump.

---

[2] After figures 17 & 15, Sugiyama, Shiotsu, Yamanaka, Hirai (2005) p 3609 & p 3610.
[3] cf Armour (2010) p 296.
[4] After figures 157 & 169, Armour (2010) p 245 & p 256.

250 grams. Basically the actuators are, besides the mechanical design, servo motors and other electric motors. The energy for jumping is stored in the flexible outer frame and can lift the robot up to a height of about seventy percent of its own.[5]

3.1.3 Modular Robots

With modular robots virtually any appearance and therefore any kind of locomotion can be achieved. It is possible, for instance, to change the robots structure from a bipedal or quadruped mode to a whole wheel that will start to roll on the ground. The next figure shows one modular robot approach that shall represent the huge variety of self-configurable robots.

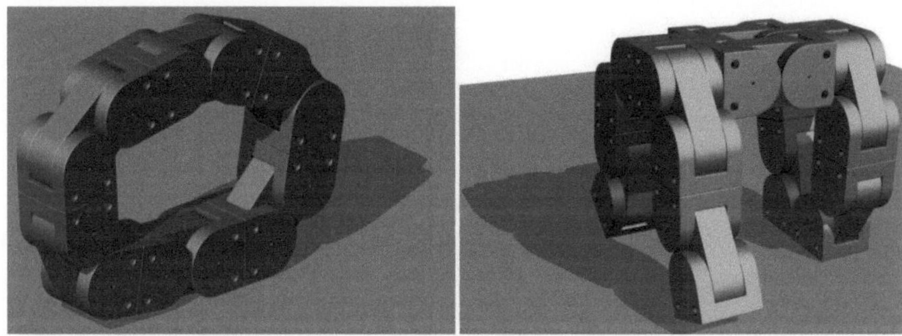

Fig. 3.3[6] *(left)* Wheel locomotion mode and walking locomotion mode *(right)* of the modular robot M-TRAN II

Further details are not being traced respective to this very idiosyncratic kind of robotics, because modular robots with their whole complexity of self-configuration would exceed the scope of this book.

3.1.4 Dynamic Rolling-Walk of a Hexapod

"ASTERISK" is the hexapod's name capable of performing a locomotion that might be considered unique among its class of robots. First of all ASTERISK is a highly symmetric hexapod to allow full movement performance in all directions including upside down operation. Various kinds of gaits were implemented into this structure. Among these is a dynamic rolling-walk allowing the robot to flip over the tips of legs. This is achieved by switching between two phases in which three and two legs would alternately have ground contact. The movement was tested under virtual conditions before implementing the program on the robot.[7]

---

[5] cf Armour (2010) p 293 ff.
[6] After figure 16, Kamimura, Kurokawa, Yoshida, Murata, Tomita, Kokaji (2005) p 322.
[7] URL: http://www-arailab.sys.es.osaka-u.ac.jp/research/ limbgroup/e_index.html. Accessed 13 Feb 2012.

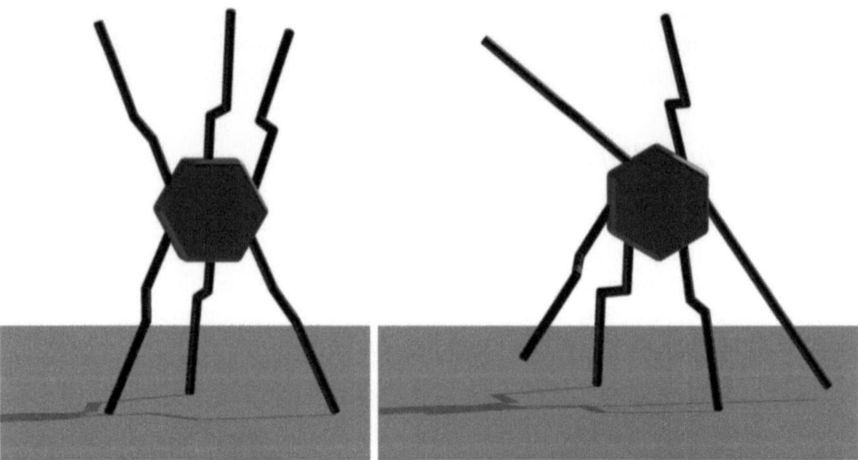

Fig. 3.4[8] Two phases of ASTERISK in which three and two legs have alternately ground contact

The rolling-walk locomotion performed by ASTERISK could be considered as a cartwheel, when thinking about the definition of locomotion that the spider presented in chapter 2.1 performs. The fact that the structure of ASTERISK is able to perform such locomotion is impressive, when considering the force the hardware has to absorb plus the energy that needs to be dedicated by the servo motors. In a video presentation of ASTERISK it can be seen that the servo motors carry the label of ROBOTIS, the company where the robot kit from the previous project came from. The servo motors seem to be those of the higher or even highest available class among the company's product line.

Questions like the one if the robot is able to take the roll-walk position without external aid remain unanswered here. Anyway, the aim is focusing more on a structure that will allow a rapid rolling movement. For further information on this it is suggested to view the document not being cited at this point: Theeravithayangkura C, Takubo T, Ohara K, Mae Y, Arai T (2009) Dynamic rolling-walk motion by limb mechanism robot ASTERISK. isbn:978-1-4244-2693-5.

3.1.5 A Walking and (Soon to Be) Rolling Robot

An interesting robot design to be mentioned was found almost at the end of the research phase. It is a six legged robot that has the structural capabilities to perform a rolling motion. The robot is able to curl up into a complete sphere, thus reminding of the project I did together with Adam Braun previous to this book.

---

[8] After figures, URL: `http://www-arailab.sys.es.osaka-u.ac.jp/research/limbgroup/e_index.html`. Accessed 13 Feb 2012.

The robotic enthusiast Kåre Halvorsen named his latest creation "MorpHex"[9]. He already designed and constructed several different robot prototypes and achieved very impressing results with them. As an advanced hobbyist in robotics he created this hexapod with six additional servo motors on the top to control the upper half of the sphere. A sooner version with a different leg structure, which was more similar to the project I did in association with Adam Braun, got discarded due to performance reasons. According to his latest blog entry[10] about the MorpHex, posted in December 2011, the rolling locomotion yet needs to be implemented. The robot performing a walking gait can already be seen in a video posted on his blog.

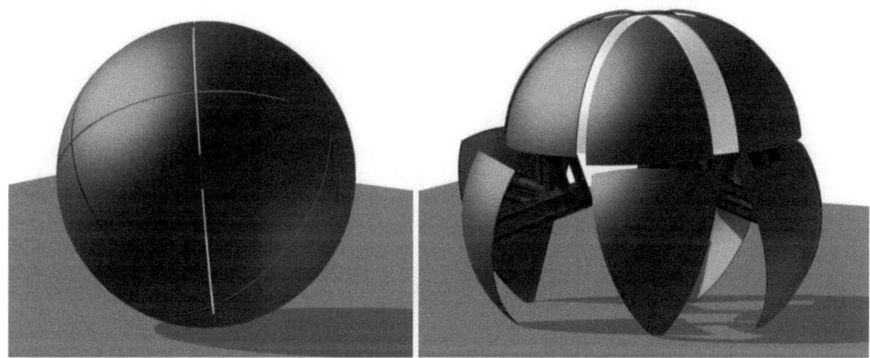

Fig. 3.5[11] *(left)* Robot in its spherical shape. *(right)* Robot is unfolded for walking.

The MorpHex shows an approach of how to combine a six legged robot with a structure for rolling locomotion purposes. The walking performance appears to be very good and undisturbed by the sphere sections mounted to the legs. If a rolling motion can be achieved with this robot still needs to be confirmed. The fact that it uses additional servo motors to control the upper half sphere parts is explained to some extent by the creator, with ideas for entertainment reasons that he would like to implement. But it is mainly because he feared a bad walking performance of the robot, if the upper parts would be mounted to the legs with additional servos as well. The fact that servo motors are added to the upper half sphere parts might also turn out beneficial when the robot performs a rolling locomotion. In case that the robot would take a condition in which the movement of the legs has no effect on locomotion, the upper parts would enable it to push itself over to a more favorable condition in which the legs can perform a propulsion movement again.

---

[9] URL: http://robot-kits.org/. Accessed 13 Feb 2012.

[10] URL: http://robot-kits.org/. Accessed 13 Feb 2012.

[11] After video, URL: http://robot-kits.org/. Accessed 13 Feb 2012.

3.1.6 Future Approaches

An example of a novel approach to realize a whole robot body with just one manufacturing process is displayed in the figure below. This structure design is based on pneumatics only, thus allowing the realization of different gaits. However, the prototype in the figure below is not able to roll. Nevertheless it is imaginable that a similar structure could curl up into a ball and that defined movements of the limbs will push the robot hence leading to a rolling locomotion. A mobile source for pressurized air would then be in need.

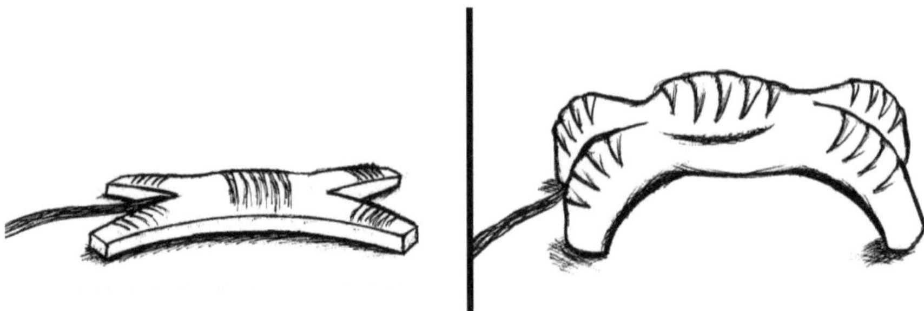

Fig. 3.6[12] Cycle of pressurization and depressurization of the soft robot, here with external air supply

3.1.7 Remark to Robotic Examples

The evaluation of all findings during the research phase was not possible due to restricted accessibility. It is therefore suggested to view the two most relevant documents out of these findings, in case of a following research after this book:

- Lianqing Yu L, Wang Y (2010) A rolling legged robot with elliptical cylinder body. isbn: 978-1-4244-7653-4
- Armour RH, Vincent JFV (2007) Rolling in Nature and Robotics: A Review. doi:10.1016/S1672-6529(07)60003-1

3.2 Actuators for a Prototype

During the project previous to this book, servo motors were used to realize a functional robot prototype. The servo motors came in a set of sensors and a microcontroller. Programming was less difficult with the provided software, compared to learning a programming language like C++. Nevertheless a brief look on other possible actuators shall be taken, each with their own advantages and disadvantages.

---

[12] After figure 2, Shepherd, Ilievski, Choi, Morin, Stokes, Mazzeo, Chen, Wang, Whitesides (2011) p 2.

3.2.1 SDM Actuators

Reducing complexity or increasing efficiency of a structure can be achieved when different materials with different properties are being combined. A metallic kitchen knife for example consists of more than just one sort of metal. Different metals and substances like carbon are brought together into an alloy. Together with an accordant tempering method, different properties like stiffness, which is good for a long lasting sharpness, flexibility that helps preventing the knife from breaking when falling on the ground and resistance against corrosion, can be combined. When two very different material properties are sought to be combined, the sandwich construction method is used. A surfboard, for example, could consist of polyurethane foam as core and an epoxy resin coat as mantle.

A novel process in combining materials is the Shape Deposition Manufacturing [SDM] method. Legs for the robot can be built by using "polymeric materials to simultaneously create the rigid links and compliant joints [...], with embedded sensing and actuation components"[13]. The figures below give an impression about what biomimetic spider legs could look like in the example of a finger from a hand gripper with its actuation parts.

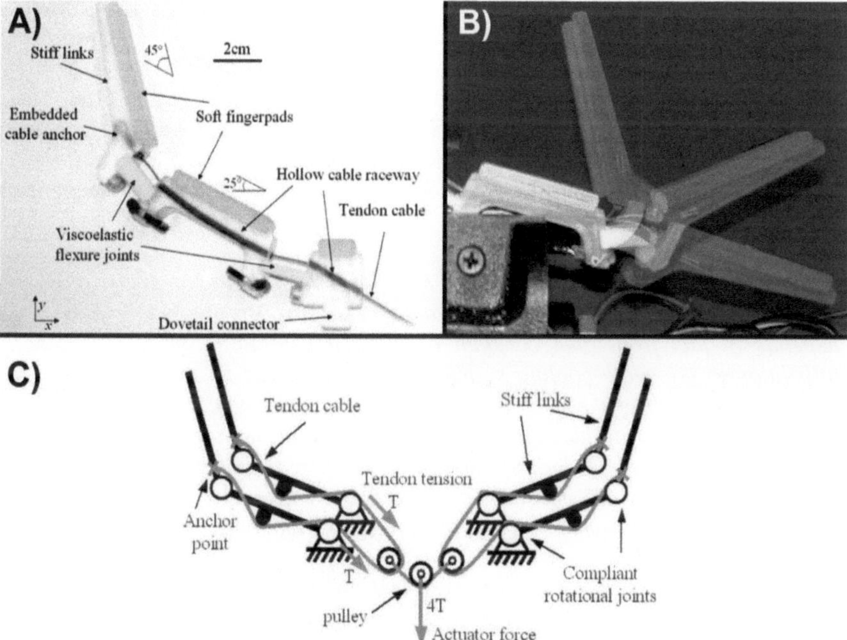

Fig 3.7[14] Finger design of a gripper hand with embedded actuation components (A). Different positions of a single link of the finger, when actuated (B). Scheme of simultaneous actuation of four fingers (C).

---

[13] Dollar, Howe (2009) p 4.
[14] cf. figures 3 & 4 & 5, Dollar, Howe (2009) p 5 ff.

An advantage would be that in case of a large number of legs on a robot the number of actuators can be kept relatively small. In case of a six-legged robot, always three legs could be actuated at the same time to perform a walking gait.

The main disadvantage in case of a robot which should walk and roll would be that a complete structure and actuation design is necessary. The engineering effort is therefore considered too high for the design thought of.

3.2.2 SMA Actuators

An older approach of using a material combination to actuate a structure is the one based on metal alloys. Electrically driven actuators are easy to use and especially servo motors are very suitable for robots in general. But for biomimetic robots with a desired 'natural' behavior, advanced properties are sought for. The "high mechanical impedance"[15] is only one of the tradeoffs that need to be taken into account when designing a model. For this reason a biomimetic robot could also have biomimetically inspired actuators. The idea of creating such kind of actuators requires the development and usage of artificial muscles. Shape memory alloys have the potential to fulfill these biomimetic requirements of artificial muscles. "SMA are alloys capable of recovering, when heated over a determined transition temperature, the large deformation impressed at low temperature."[16] Dependent on the actual combination of materials, an SMA can change its form with an increasing and/or lowering temperature. This behavior is called the Shape Memory Effect due to the fact that the material needs to memorize its shape under certain temperatures combined with mechanical training that simulates the wanted condition. The SME is not erratic in SMA and thus a smooth movement in an actuator can be achieved.

"On the other hand, a control variable such as temperature strongly affects the ultimate performances of SMA actuators. The rate of obtainable movements depends in fact from time constant of thermal transient. For this reason the utilization of adequate cooling systems is often required in order to achieve the desired bandwidth of the actuation system. Since SMA actuators can be seen as thermal machines, producing mechanical work from heating, their efficiency is extremely low: usually it does not exceed 10%. Then, from a general point of view with respect to existing conventional motors, SMA actuators must be considered as a complementary technology rather than a substitute one."[17]

This statement still holds true in a sense that SMA are used for a huge variety of applications[18], but in robotics they are not established yet as effective actuators. Therefore I will also refrain from developing a SMA suitable actuator structure.

---

[15] Pratt (2003) p 195.

[16] Bergamasco, Salsedo, Dario (1993) p 508.

[17] Bergamasco, Salsedo, Dario (1993) p 514.

[18] Kumar, Lagoudas (2008) p 29 ff.

3.2.3 EAP Actuators

Another approach to create artificial muscles is the one of electroactive polymers (EAPs). These polymers constitute a combination of different materials.

"These materials have functional similarities to biological muscles, including resilience, damage tolerance, and large actuation strains (stretching, contracting, or bending). EAP-based actuators may be used to eliminate the need for gears, bearings, and other components that complicate the construction of robots and are expensive, heavy and fail prematurely. Visco-elastic EAP materials can potentially provide more lifelike aesthetics, vibration and shock dampening, and more flexible actuator configurations. Exploiting these properties may lead to the development of artificial muscles that may be applied to mimic the movements of animals and insects, and even enable the movement of the covering skin to define the character of the robots and provide expressivity."[19]

Regarding to this statement, an ideal technology for artificial muscles is found for this project. But today the development of EAPs seems still to be far away from a product ready for the market and thus they are not easily available for the creation of a biomimetic robot. In fact EAPs are still a matter of laboratory tests and in contrast to the statement cited above there are a lot of issues regarding reliability, strength, and many other parameters. So far EAPs have a certain potential and a huge variety of different material combination approaches is under research, but an outstanding breakthrough is still pending.[20]

3.2.4 Fluidics - Hydraulics

A manifest actuator structure is the one used by the spider itself. Spiders have a system based on hydraulically pressurized limbs combined with muscles. The pump to generate the pressure is located within the spider's body. Pressure forces the legs to stretch and muscles guide their movement. Valves fulfilling a task like the ones used in the automation industry do were not discovered to exist in the spider so far. It is suggested that the pressure is distributed by a built-in structure of the legs, interacting with the muscles and joints.[21] The effect of hydraulically stretched legs can easily be observed on dead spiders, which have curled up legs resulting from the missing pressure of the pump. Using a hydraulic system in the same way the spider does is still complicated and needs further investigation. Additionally hydraulics is mostly used for actuators that need to transmit high pressures like in shovel dozers and excavators. A robotic example with hydraulic legs is the ALDURO[22], which is capable of carrying a human being.

---

[19] Meijer, Bar-Cohen, Full (2003) p 31 f.
[20] cf. Choi, Jung, Koo, Nam (2007) p 49 ff.
[21] cf. Blickhan, Petkun, Weihmann, Karner (2005) p 21 ff.
[22] Hiller (2005) p 191.

3.2.5 Fluidics - Pneumatics

Another sector of fluidics is pneumatics. Instead of a liquid medium a gaseous medium is compressed and used to transmit energy. The use of pneumatics is less complex because air can be used as a medium instead of a fluid. In terms of biomimetic applications pneumatics have some interesting properties.

"Pneumatic actuators have biological characteristics and certain very attractive engineering characteristics as well. Like muscle, a pneumatic actuator typically increases its stiffness as force is increased. Pneumatics are used extensively in the factory automation industry, because of their high-power density and specific power, high-force density and specific force, their lack of need for transmissions, their ease of control for digital full motion tasks by digital valves, and the fact that fluid (air) leaks are benign. Furthermore by regulating the pressure across a pneumatic actuator, one can roughly control its output force."[23]

Furthermore, the usage of pneumatics is not limited to simple cylinders. The German company FESTO, which is specialized in the field of pneumatics for industrial purposes, brought the so called 'fluidic' or 'pneumatic muscle' to the market.[24] Numerous other biomimetic inventions and robots were developed and built by FESTO and its partners. The Fraunhofer IPA, which worked together with FESTO on a project before, now developed a biomimetic spider robot and presented it on the exhibition EUROMOLD 2011 in Frankfurt, Germany. The walking gait of the spider is completely achieved with pneumatics. Applying pressured air to the bellows of the robot, which are located on its leg joints and inside the body, leads to stretching and contraction of the legs.

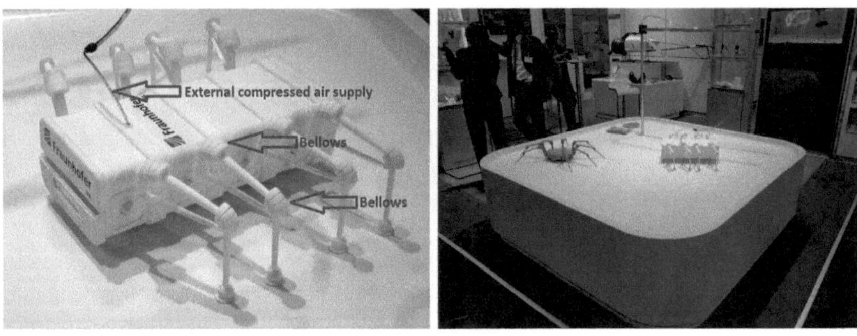

Fig. 3.8 *(left)* First biomimetic spider robot design with bellows by Fraunhofer IPA. *(right)* Relative size of the first and second prototype. First prototype connected to an external air compressor.

---

[23] Pratt (2003) p 192.
[24] cf. Nachtigall (2002) p 180.

The two robot prototypes are produced with the SLS method, a rapid prototype and rapid manufacturing process. The bellows consist completely of the SLS material. Functionality of the bellows can be calculated prior to the manufacturing process and therefore be changed for different loads which the robot needs to carry.

Fig. 3.9 *(left & middle)* Second spider robot prototype by Fraunhofer IPA. *(right)* Close-up view of the bellows inside the robot housing.

Four of the legs move diagonally at the same time. Two push and two pull, while the other four legs are lifted up. Along with the bellows some controller-regulated valves are needed. These prototypes rely on an external compressor as source for pressurized air, which later is thought to be replaced by a built-in compressor.

The autarkical supply with pressurized air is essential for the robot implementation that is sought in this book. The reason for this is that a rolling robot's independency would significantly deteriorate if it needs to be connected to an external source for pressurized air. Efficient, in terms of size and energy consumption, on-board air compressors need to be used, which require a power source like a battery, for instance, to drive them. Hence it is questionable if a mobile pneumatic robot can compete with one that converts the energy from the autarkical source directly into a movement, like a robot with servo motors, for instance.

3.2.6 Mechanical Solution

With an intelligent mechanical design, certain active actuators could be replaced by a passive structure. For walking purposes the legs of the robot would perform a parallel movement to a certain point. This could be achieved with a structure as shown in the figure below.

Mechanics does not just offer solutions to save actuators, but also offers the possibility to bring in parameters like elasticity to the system, which is not given when, for example, servo motor actuators come to use. The disadvantage of these

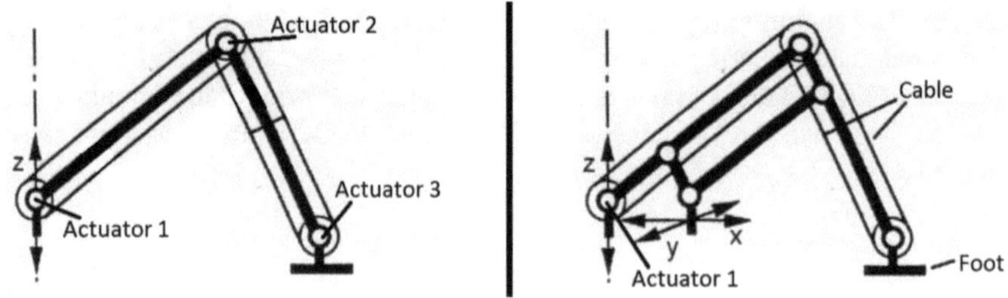

Fig 3.10[25] *(left)* Robot leg with three actuators. *(right)* Leg with mechanical solution to save two actuators.

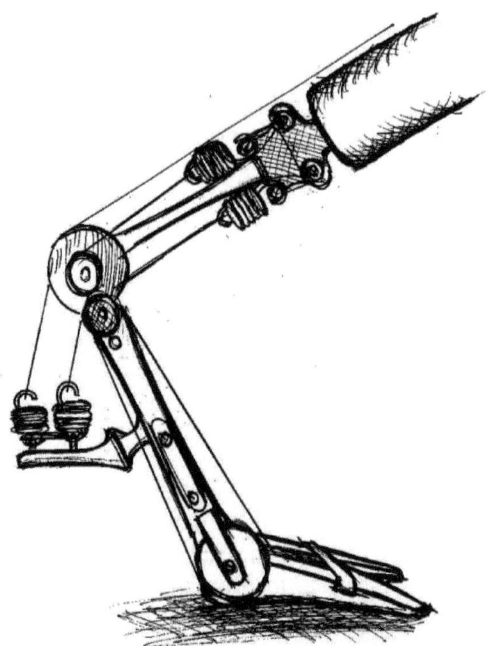

Fig. 3.11[26] Series-elastic actuators

elastic actuators is that they are hard to control and a given structure cannot be changed or adjusted as fast as it is possible with servo motors, which can be controlled with programming.

In reference to a robot that will eventually curl up its legs, the system becomes far too complicated for this purpose - even though a spring system, to support the servo motors dealing with the high loads in walking mode, was thought to be used in the previous project.

---

[25] cf. figure 10-10 B, Nachtigall (2002) p 184, figure changed.
[26] After figure 7.12, Pratt (2003) p 198.

3.2.7 Mechatronics

As seen before in some robot examples, mechatronic components like servo motors still play a major role in robotics. In terms of handling and use they are easy to control and program without undesired dynamical behavior, when damped accordingly in the settings through programming. Limits can be, or are, defined by clear mechanical constraints such as maximum torque and operating temperature, for instance. The use of dynamical systems is limited, because no energy can be stored within the electrical motor like in a spring. Additional electronics would be necessary to recuperate energy through servo motors from a dampening movement. This is quite complicated compared to a mechanical solution where dampening energy could easily be stored in a spring or a pneumatic cylinder that would then have a preload due to increased pressure. Despite that, servo motors are technically not well-suited to work as generators. In fact the power consumption is very high. But still the performance in general and its availability in a huge number of sizes and other parameters make this mechatronic component become an easy to use technical joint. A short comparison of three of the above listed actuator technologies, in terms of controllability and flexibility in construction, for instance, can be seen in the following figure. This figure will also back up the decision to use the robot kit from the previous project again, when considering the flexibility in construction as the most important factor, which is due to the reason that a robot prototype needs to be developed, tested and adjusted.

	Transmission method		
Assessment criteria	mechanical	hydraulic	electric
force density	good	okay	bad
transmittable over middle distances	bad	okay	good
controllability	bad	okay	okay
efficiency factor	okay	bad	bad
flexibility in construction, supply of components	bad (expensive)	okay (expensive)	good (inexpensive)

Fig. 3.12[27] Comparison of force and energy transmission in different systems

---

[27] After figure 1.4-1, Murrenhof (2007) p 16.

3.2.8 Robot Kit

For this project the 'BIOLOID Premium kit' from the company ROBOTIS, with headquarters located in Seoul, South Korea, is used. The kit comes with an own software tool for programming, which does not necessarily require the comprehensive knowledge of a programming language like C++. It is designed in a way to allow a more intuitive programming. However, knowing a programming language is thought to be advantageous.

Fig. 3.13 BIOLOID Premium Kit, parts and hexapod example

The robot kit is mainly designed to build a humanoid robot with the option to choose between three different configurations. These three types of humanoid robots can be found in the manuals included in the kit. Despite that, twenty-six other examples in three levels, from beginner through intermediate up to advanced, can be built. The manuals for the other robots are provided on the manufacturers' website 'www.robotis.com'. An advanced example is a hexapod called the 'King Spider', which can be seen in the figure above.

The kit contains eighteen servo motors that can be configured in a full rotational wheel mode as well. Signals and power is transmitted to the servos through a network wiring. Each servo therefore has an own ID-number. Beside the fine adjustable movement range of 300°, the servo motors can be controlled with regard to values like temperature, torque, speed, load and softness in joint mode. A microcontroller, battery-pack and charging device is provided as well. The kit also contains two infrared sensors, a distance measurement sensor as well as a gyro sensor module with two sensed axes. The robot can optionally be controlled with the provided remote control through an infrared receiver.

Programming of the microcontroller is done with several software applications that are provided on the CD included in the kit. The latest versions of the software as well as updates for the microcontroller are available for download on the website mentioned. The microcontroller will be connected to the computer through an adapter, which needs to be switched to the desired purpose of sending programs to the microcontroller.

The 'RoboPlus' interface permits access to the three programs:

-    'RoboPlus Task'; programing functions like IF, WHILE and CALL.
-    'RoboPlus Manager'; setting the servos to joint or wheel mode for instance.
-    'RoboPlus Motion'; defining position, speed and other parameters of the servos.

```
RoboPlus Task - bio_prm_quadrupedrobot_en
File(F)   Edit(E)   Program(P)   Tool(T)   Help(H)
                                    Controller: CM-510      ▼   Port: COM3        ▼
 1   // Bioloid Premium Kit Quadruped Walking Robot Task code
 2   // Initial Release : July 9, 2010.
 3   // Revised : Assembly check added (September 28, 2010).
 4   // How the program works
 5   // The robot avoids obstacles on its path.
 6   START PROGRAM
 7   {
 8       IF ( Button == D )
 9           JUMP AssemblyCheckMode
10       LOOP FOR ( ID = 1 ~ 8 )
11       {
12           IF ( ID[ID]: ADDR[3(b)] != ID )
13               CALL AssemblyError
14           // Dynamixel ID check
15           IF ( ID[ID]: ADDR[8(w)] == 0 )
16               ID[ID]: ADDR[8(w)] = 1023
17           // Dynamixel mode check (joint or wheel)
18       }
19       // Assembly check
20       CALL Initial
21
22       ENDLESS LOOP
23       {
24           IF ( PORT[1] >= 250 )
25           {
26               CALL Stop
27               CALL GoRight
28           }
29           IF ( PORT[2] >= 40 )
30           {
Ready
```

Fig. 3.14 Screenshot of the RoboPlus Task program showing provided quadruped task code example

With the 'RoboPlus Task' program the 'intelligence' of the robot is programmed. Typical functions can be generated with AND, WHILE, IF, LOOP and several other commands. Sensor values and even single servo parameters can be defined here. Unfortunately functions and values are only to be set from a drop down menu. On the one hand this reduces the probability for mistakes. On the other hand it slows down the process of programming drastically, because sometimes a mouse button double click needs to be performed three times before the desired value can be changed. For professionals there is the option to program the robot entirely with the language C. This procedure is supported by the software.

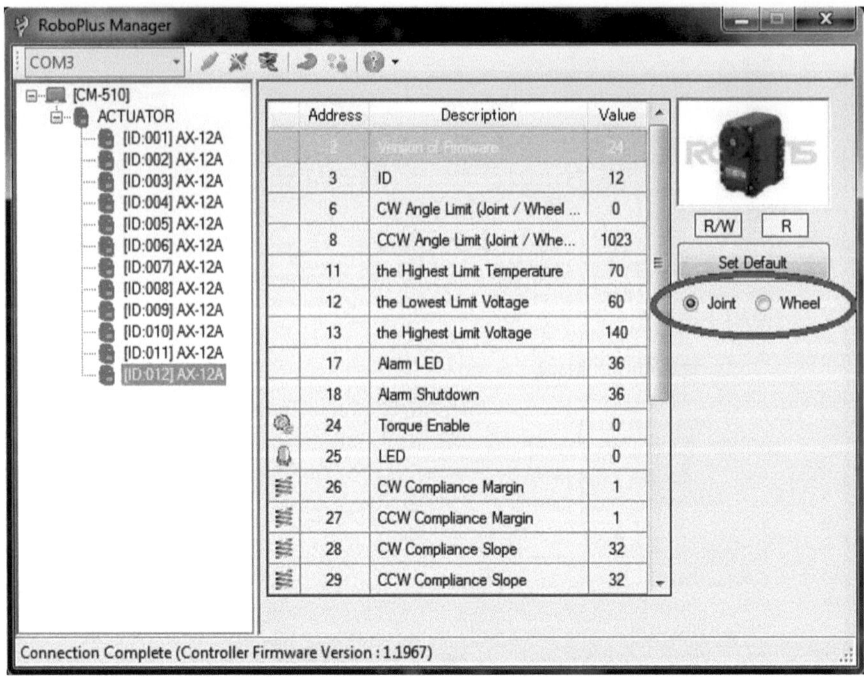

Fig. 3.15 Screenshot of the RoboPlus Manager with connected microcontroller and twelve available servo motors. The option to change the servo motor number twelve into wheel mode is highlighted.

Hardware settings can be changed with the 'RoboPlus Manager'. The most useful feature thereby is the option to set the servo from joint mode to wheel mode in which the servo motor would act like a normal electrical motor. Changing other parameters than this was not tried out so far. Limiting the torque for example, to a lower value than the default one, is of no use for this project. If possible, changing the maximum temperature within the servo to a higher level, would surely lead to a hardware malfunction. The servos are set to a default level were overheating and torque exceeding will not lead to defective hardware. Programming mistakes are tolerated by the default settings of the hardware. One example that can be mentioned here is when the servo turns into the wrong direction when working with full speed - it might crash into its own base-joint. The LED on the servo would blink in red color and the torque would immediately turn off to prevent failure. Switching the microcontroller off and on again, would delete the error.

Motion sequences can be generated with the 'RoboPlus Motion' program. The servos revolution angle of 300° can be achieved by steps from zero to 1023, while 512 is the initial position. Every line [as seen in the figure above, on the left hand

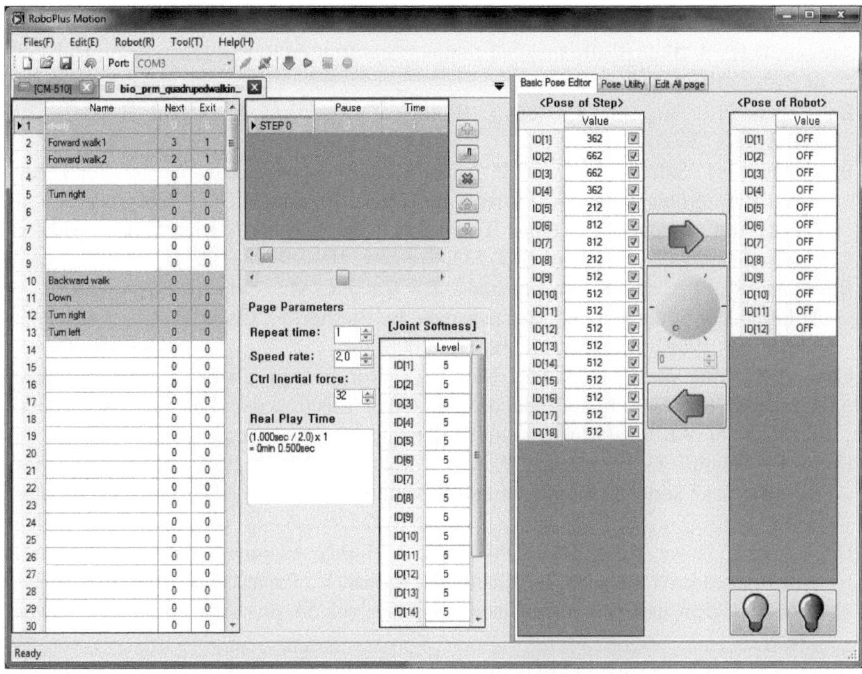

Fig. 3.16 Screenshot of the RoboPlus Motion program. The microcontroller is connected and the quadruped motion sample is loaded. Certain tools are available to set speed and position of each single actuator.

side in green] can be named at the user's own choice and contain a maximum of seven steps [from zero to six, as seen in the middle with STEP 0 in blue]. In case seven steps for a motion are not sufficient, which is most likely the case, then the next line with another seven steps will be used. Later, the task program will call a motion line by its number and automatically execute the following lines when set to do so. A very useful setting is the alternative to switch torque on and off for an individual actuator, for multiple ones or for all. The buttons for that are represented by the two light bulbs in the bottom right corner of Fig 3.16. This enables the user to move the actuators into a desired position when torque is switched off. Switching torque back on will show the position of each actuator in the list [right hand side of Fig 3.16]. These values can be assigned to a step with the green arrows. Setting up rough motions becomes very easy with this teach-in method. To set up precise motions it is still necessary to adjust the rough setting manually.

References

Armour, R.H.: A Biologically Inspired Jumping and Rolling Robot. Dissertation University of Bath (2010)

Bar-Cohen, Y., Breazeal, C. (eds.): Biologically Inspired Intelligent Robots. SPIE, Bellingham (2003)

Bergamasco, M., Salsedo, F., Dario, P.: Shape Memory Alloy Linear Actuators for Tendon-Based Biomorphic Actuating Systems. In: Dario, P., Sandini, G., Aebischer, P. (eds.) Robots and Biological Systems: Towards a New Bionics? Series F: Computer and System Sciences, vol. 102, pp. 507–533. Springer, Heidelberg (1993)

Blickhan, R., Petkun, S., Weihmann, T., Karner, M.: Schnelle Bewegungen bei Arthropoden: Strategien und Mechanismen. In: Pfeiffer, F., Cruse, H. (eds.) Autonomes Laufen, pp. 19–42. Springer, Heidelberg (2005)

Choi, H.R., Jung, K.M., Koo, J.C., Nam, J.D.: Robot Applications of Artificial Muscle Actuators. In: Kim, K.J., Tadokoro, S. (eds.) Electroactive Polymers for Robotics Applications: Artificial Muscles and Sensors, pp. 49–89. Springer, London (2007)

Dario, P., Sandini, G., Aebischer, P. (eds.): Robots and Biological Systems: Towards a New Bionics? Series F: Computer and System Sciences, vol. 102. Springer, Heidelberg (1993)

Dollar, A.M., Howe, R.D.: The SDM Hand: A Highly Adaptive Compliant Grasper for Unstructured Environments. In: Khatib, O., Kumar, V., Pappas, G.J. (eds.) Experimental Robotics. Springer Tracts in Advanced Robotics, vol. 54, pp. 3–11. Springer, Heidelberg (2009)

Hiller, M.: Autonomes hydraulisch angetriebenes Schreitfahrwerk ALDURO. In: Pfeiffer, F., Cruse, H. (eds.) Autonomes Laufen, pp. 191–198. Springer, Heidelberg (2005)

Kamimura, A., Kurokawa, H., Yoshida, E., Murata, S., Tomita, K., Kokaji, S.: Automatic Locomotion Design and Experiments for a Modular Robotic System (2005), http://unit.aist.go.jp/is/frrg/dsysd/mtran3/papers/T-Mech05.pdf (accessed February 13, 2012)

Khatib, O., Kumar, V., Pappas, G. (eds.): Experimental Robotics: The Eleventh International Symposium, vol. 54. Springer, Heidelberg (2009)

Kim, K.J., Tadokoro, S. (eds.): Electroactive Polymers for Robotics Applications: Artificial Muscles and Sensors. Springer, London (2007)

Kumar, P.K., Lagoudas, D.C.: Introduction to Shape Memory Alloys. In: Lagoudas, D.C. (ed.) Shape Memory Alloys: Modeling and Engineering Applications. Springer, New York (2008)

Lagoudas, D.C. (ed.): Shape Memory Alloys: Modeling and Engineering Applications. Springer, New York (2008)

Meijer, K., Bar-Cohen, Y., Full, R.J.: Biological Inspiration for Musclelike Actuators of Robots. In: Bar-Cohen, Y., Breazeal, C. (eds.) Biologically Inspired Intelligent Robots, pp. 26–41. SPIE, Bellingham (2003)

Murrenhoff, H.: Grundlagen der Fluidtechnik: Teil 1: Hydraulik, 5th edn. Shaker, Aachen (2007)

Nachtigall, W.: Bionik: Grundlagen und Beispiele für Ingenieure und Naturwissenschaftler, 2nd edn. Springer, Heidelberg (2002)

Pfeiffer, F., Cruse, H. (eds.): Autonomes Laufen. Springer, Heidelberg (2005)

Pratt, G.A.: Biologically Inspired Components of Robots-Sensors, Actuators, and Power Supplies. In: Bar-Cohen, Y., Breazeal, C. (eds.) Biologically Inspired Intelligent Robots, pp. 179–202. SPIE, Bellingham (2003)

Shepherd, R.F., Ilievski, F., Choi, W., Morin, S.A., Stokes, A.A., Mazzeo, A.D., Chen, X., Wang, M., Whitesides, G.M.: Multigait soft robot (2011),
http://gmwgroup.harvard.edu/pubs/pdf/1135.pdf
(accessed February 13, 2012)

Sugiyama, Y., Shiotsu, A., Yamanaka, M., Hirai, S.: Circular/Spherical Robots for Crawling and Jumping (2005),
http://www.mint.se.ritsumei.ac.jp/articles/05/
ICRA2005sugiyama.pdf (accessed February 13, 2012)

http://www-arailab.sys.es.osaka-u.ac.jp/research/limbgroup/
e_index.html (accessed February 13, 2012)

http://robot-kits.org/ (accessed February 13, 2012)

4 Biomimetically Inspired Robot Prototype

The aim of this book also focuses on keeping the biomimetic transformation on an applicable level. The reasons for this are limited resources as well as given time constraints. To analyze the applicability of different concepts and setting up own approaches, an analysis will be applied for the classification of different robot approaches.

4.1 Parameters

The classification table that comes to use consists of two axes that allow the free combination of two parameters. In this case, the parameters need to be defined first. Existing robot approaches are sorted along these two axes to the box matching best within the table. After that, own suggestions complete the table and result in parameters with values that are to be obtained in the further work.

Some attributes can be defined and easily assigned to technical samples when thinking about a spider. A spider has a metabolism to 'run its system'. It needs to eat and digest. A robot 'consumes' electric energy, for instance, that is provided by an accumulator or an external energy source. The fact that the robot is dependent on an external source to charge when doing so, is disregarded here. The topic of supplying energy would be another, however important question, when the robot is suggested to operate in an autarkical way. A spider has eyes, hair and other mechanisms to recognize its environment. This can be achieved to a certain degree using various kinds of sensors, such as the distance measurement sensor and light recognition sensor, for example. The spider also has legs, for walking, jumping and in the case under special observation here, for rolling locomotion. The legs show a wide range of flexible movements.

The parameters for the classification table should be based upon non-exchangeable attributes. The power source for instance is an exchangeable attribute. In case of the use of servo motors, it is irrelevant from which power source the electrical energy will come from. The sensors as well are exchangeable. Some basic sensors will surely enhance locomotion performance and capability of the robot, but functionality can also be granted with well-defined programming. The rolling robot from the previous project for example, was built without any sensors.

The components that most certainly will have an effect on the performance are the legs. The spider has not any less than eight of these. But are really eight legs

Ralf Simon King: BiLBIQ: A Biologically Inspired Robot, BIOSYSROB 2, pp. 49–76.
DOI: 10.1007/ 978-3-642-34682-8_4   © Springer-Verlag Berlin Heidelberg 2013

needed, or will a few legs less do the job as well? This is of course not to be found out on the biological sample, but on the robot. And hence the first parameter and axis in the classification table should be – the **number of legs**.

Another key parameter seems to be the agility of the legs. Folding them to form a wheel or a sphere is not possible with a stiff leg consisting of one link only. The spider leg is composed of several links. But the leg cannot be understood by comparing it to the one of a human. While a human being has only muscles to contract in one and stretch in the opposite direction the spider additionally preloads its legs hydraulically. The links of the spider are connected through the joints and therefore mostly perform a similar movement at a time. Nevertheless, a leg of the spider can bend, stretch, twist and so on, in several directions like a human leg can. In how many directions a single link can move is determined by the DOF it has. That means, to twist a link around its own axis, one DOF is needed. To lift it up and down around a joint, another DOF is needed. The question now is: How many DOFs does one leg need, to fulfill the performance sought for? Hence the second parameter will be – the **DOF per leg**.

4.2   Robot Classification Table

The classification of the robot examples is based on the assumption that each single actuator of them equals one DOF: The Jollbot [A-3] is considered having three DOF because it can move freely in all directions of the coordinate system. It is capable of moving planar into the x- and y-direction and, when jumping, it even moves into the z-direction. The robot of Kåre Halvorsen [E-4 & G-2] can be considered as a hexapod or twelve-legged robot. It has twenty-five actuators: One in the center of the robot, eighteen on the legs for walking and six on the upper half. It has three DOF per walking leg, but the additional seven actuators need to be taken into account as well. For this reason the robot is shown twice in the table. The robot in section G-1 is able to flip itself over by ascending and retracting the one DOF actuators in a certain moving pattern.

Now that the classification table contains some robot examples to exemplify what kind of structures can be achieved, a new robot structure can be determined by combining the two parameters in a new way: The boxes highlighted in brown exhibit configurations that will not lead to a structure which combines walking and rolling locomotion. This is due to the fact that a walking locomotion cannot be achieved with a monopod and the combination of a large number of legs with a very limited amount of DOF will not lead to the sought structure as well. The boxes highlighted in blue are considered to contain an unnecessary large amount of DOF and/or legs. As mentioned earlier I want to keep the robot design on a manageable level. Therefore the number of actuators should be kept low. The boxes highlighted in green exhibit the field for approaches with a high potential to fulfill the attributes that will be defined more precisely in the following chapter.

The classification of robots with the help of a table is a simple approach to estimate the minimal effort that needs to be put in the development of a structure that is designed for walking and rolling locomotion.

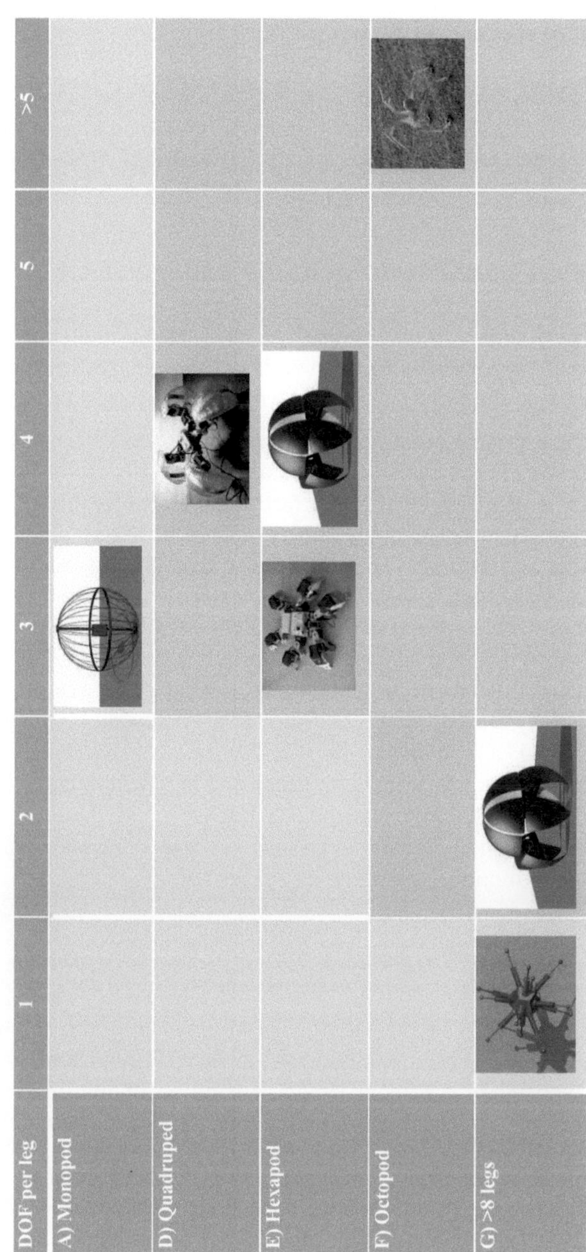

Table 4.1[1] General ranking of different robot approaches to exemplify the DOF per leg/amount of legs correlation

---

[1] **A-3:** after figure 157, Armour (2010) p 245 **D-4**: Quadruped from the previous project **E-3:** Hexapod 'King Spider' built in the previous project **E-4 & G-2:** after video, URL: `http://robot-kits.org/` **F->5**: cf figure Rechenberg (2012) PowerPoint **G-1:** after figure 1, Geheorghe, Alexandrescu, Duminică (2011) p 18.

4.3 Performance Measures

For a meaningful implementation and reduction of the topic's complexity, some key parameters need to be defined. Principles of the biological archetype are going to be extracted and an enhancement is evaluated with respect to additional properties.

4.3.1 Main Characteristics of the Biological Archetype

As mentioned earlier in the comparison chart of the two spider types *Cebrennus villosus* and *Carparachne aureoflava*, the *Cebrennus* performs a somersault in the air in contrast to the *Carparachne*, which rather performs a passive cartwheel.

- The main principle obeyed by both spiders is that they roll by rotating their whole body but not around a fixed axle.
- The main attention lies on the fact that the *Cebrennus* spider uses the same structure, namely its legs, for two very different kinds of locomotion – walking and rolling. The *Cebrennus* can do this on its own, while the *Carparachne* needs to make use of external aid, namely gravity force, and hence needs a slope to start rolling after curling up.
- Performing a somersault in the air or a handspring implies by definition that the main body has no ground contact during all phases of the rolling motion. This is why the *Cebrennus* does not do a somersault on the ground.
- Performing a somersault in the air or a handspring also implies that extremities, such as arms with hands and legs with feet, or just legs in case of a spider, are used as propulsion actuators.
- These propulsion actuators can also serve as controllers for the direction of the rolling motion.

4.3.2 Additional Properties

As seen among some biological archetypes in chapter two, animals mostly roll up as a defense mechanism. A defense or rather protection mechanism for the robot could be realized with a leg structure as featured by the robot from the previous project. Legs that are used for locomotion as well as they serve as a shell when curled up could protect the robot against environmental disturbances that are harmful to the more fragile mechatronic components such as servo motors and sensors. In case of an environmentally investigating robot, the shell could prevent the robot from damage caused by bad weather like heavy rain or stormy wind with flying debris. Of course the robot needs to take a safe position or stand to not be blown away like the tumbleweed.

Regardless of how beneficial this additional property would be for the robot, it is refrained to implement it into the following prototype. One reason is that the spider has no additional protection shield either. Another reason is the bad walking performance of the robot from the previous project that had a full sphere protection shield. So it is for the robots performance sake that it will not exhibit this additional parameter at the moment.

4.4 Creation of the Physical Robot Prototype

4.4.1 Merge and Derive Theory

Finally theory needs to be applied onto a physical approach. The physical performance of the robot can only be estimated by now. Therefore a final decision needs to be made with respect to the parameters that should be obeyed and to which hardware components can be used. In the introduction it was concluded that the robot from the previous project delivered promising results and suggestions were made on how the structure could be enhanced. Thanks to this experience it can be derived that the combination of a suitable servo motor alignment with a newly developed leg design will lead to a successful structure and thus a successful implementation. Hence, the first consideration is:

- **Hardware configuration and leg design**

The spider *Cebrennus villosus* is a self-contained, highly sophisticated organic system that can realize two very different ways of locomotion. It uses the same structure for both ways of movement and due to the fact that an ineffective behavior is not tolerated by evolution through selection, the spider cannot afford wasting precious resources. Hence every part of the spider has to fulfill a purpose in both kinds of locomotion - walking and rolling. Some parts of the spider's body might play a more important role than others, of course, but the quintessence is that all parts interact in some way at least. The fact that the spider is a self-contained system interacting with itself, implies that the robot should be a closed system as well, merging the most important parameters to allow for a biomimetic implementation. Focusing on a single type of locomotion and disregarding the other might most likely lead to the path of an improper implementation and most certainly away from the biological archetype. Hence the second consideration is:

- **Examine the whole system - implement the whole system**

The classification table exemplifies where the development of a robot configuration can be started. It is not desired to exactly copy the spider and construct an octopod into which the locomotion could be implemented both easily and successfully [what finally cannot be guaranteed anyway]. Rather than that, a design is sought fulfilling the set of parameters and disregarding unnecessary additional structure - especially additional actuators. The idea of an octopod as a robotic approach toward mimicking nature's blueprint can easily be rejected when taking into consideration that the legs of the robot can be designed freely. The leg design is thought to replace some pairs of legs in contrast to the spider, which needs to form an abstract sphere, or rather a double wheel, with its legs for the above mentioned reasons. If the spider would have some additional leg pairs this would result in a more circular shape when curled up. Robots' legs, however, simply need to be shaped like a circle, half circle or one-third circle depending on the number of legs.

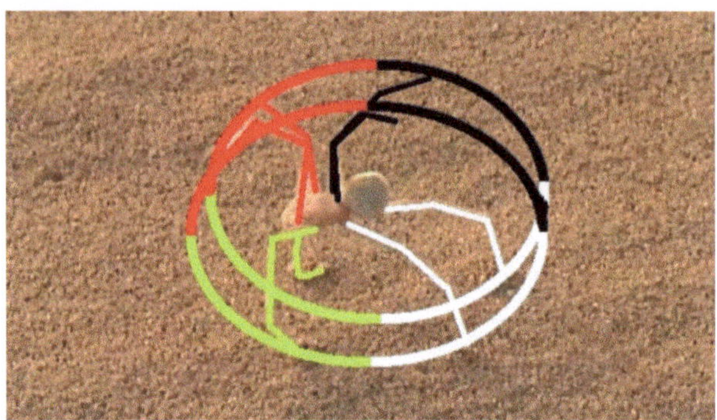

Fig. 4.1[2] Visualization of the *Cebrennus villosus* forming two imaginary abstract circles, each subdivided into four parts representing the according legs

Taking a look back on the classification table the application of four or six legs appears to be the most reasonable approach. First of all, an uneven number of legs is disregarded entirely because evolution favors symmetric designs [there are scorpions and other animals, which are mostly akin with spiders, that have evolved different sized claws, for instance, but the total amount of extremities always equals an even number]. Additionally, in case of a tripod, a stable non-dynamical walking gait cannot be achieved, because at least three legs are needed to guarantee a safe stand of the robot. Furthermore, there is no apparent advantage of a five- or seven-legged robot, especially because a robot with two axially symmetrical sides, in contrast, is expected to perform better when fulfilling a somersault to the front.

A hexapod offers the advantage of rapid movement with three legs resting on the ground to ensure a safe stand and three legs that proceed into the walking direction. In case of developing a sphere- or ring-like structure with the robot's legs, the ring or sphere parts can be kept relatively small, because the shell can be divided into more individual parts. It is also more likely that the robot, when curled up, will take a good initial position to trigger the rolling locomotion due to the fact that it is unlikely that the robot will remain exactly on one third of its circularly shaped legs, thus bringing itself into an unstable position. But there is an important disadvantage about this. Knowing that one leg of the robot needs to have at least two DOF and hence two actuators only for walking purposes then six additional actuators are needed, compared to a quadruped. This is because at least one more actuator is necessary to position the leg into the rolling configuration. This does not only affect the robot's weight, a fact that might lead to a failure of the hardware that is available for this project due to higher torques, but also the

---

[2] cf. figure Rechenberg (2012) PowerPoint.

complexity of controlling and programming the system. The fact that more actuators might consume more energy is disregarded, because it might also be the case that more actuators need to move less to achieve the same results in terms of time and distance.

A quadruped, in contrast, is disadvantageous because it can only lift one leg at a time when it is supposed to always ensure a safe tripod stand. Nevertheless, besides the positive effect of decreased control effort, the probability of interference between the legs in the rolling mode is lower. Additionally the scope for an actuator alignment and leg design is considered to be higher, because the structure allows for an easier symmetrical actuators assembly.

4.4.2   The Quadruped

Being endowed with the quadruped experience from the previous project and the important benefit regarding the bigger scope due to a new leg design and actuator alignment, the quadruped approach is going to be traced further. While keeping in mind that the robot needs to perform somersaults in a later state, an alignment was designed that still should enable the robot to walk properly. Since the quadruped from the previous project failed in performing a walking gait, the first robotic approach in this project focuses slightly more on this particular issue rather than on the one of a rolling locomotion. The very first quadruped design builds up on the idea of aligning the leg actuators as closely together as possible. In contrast to the 'SpiderBall' [name of the robot from the previous project], three instead of four actuators are aligned to form one leg. The leg alignment of the two external actuators was changed as well to provide the robot with a larger movement range thus enabling it to take larger steps.

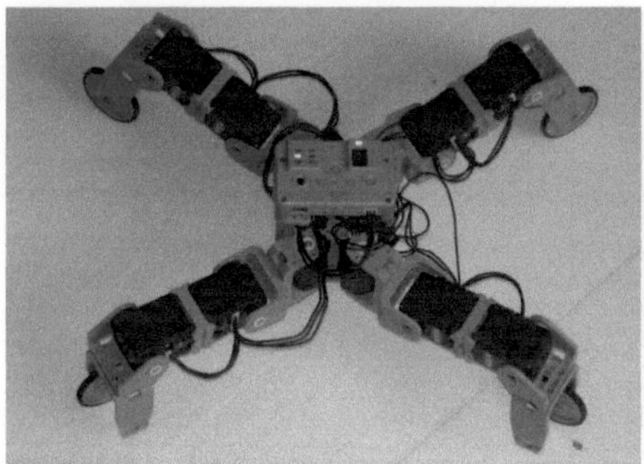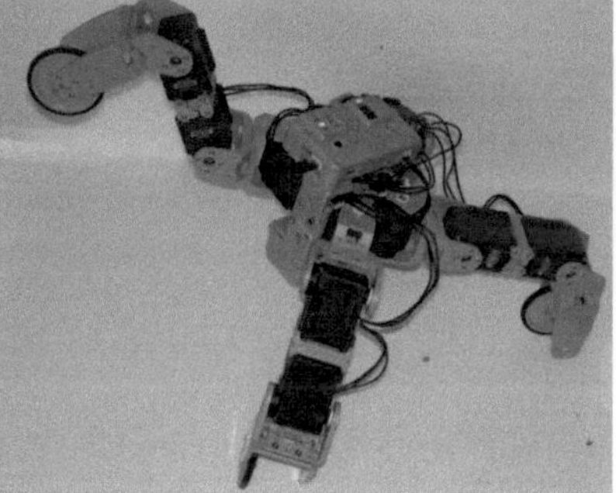

Fig. 4.2 First quadruped design with compactly arranged actuators as body and 'long' legs

During the first manual test of the robot's movement range it turned out instantly that the internal, closely packed actuators interfere with each other and hence limit their own movement range when walking and curling up into the desired rolling shape. This circumstance was not foreseen, since the SpiderBall had closely packed actuators as well. What has changed is that the new quadruped has a mirror-symmetrical alignment of actuators as opposed to the SpiderBall's irregular cross shaped alignment of actuators. Hence, the second robot is built with additional parts from the robot kit to design a 'torso' for the robot. Another convenient feature of this new design is that the accumulator can be placed space-saving and safely inside the torso structure and the microcontroller can be mounted thoroughly on top of it.

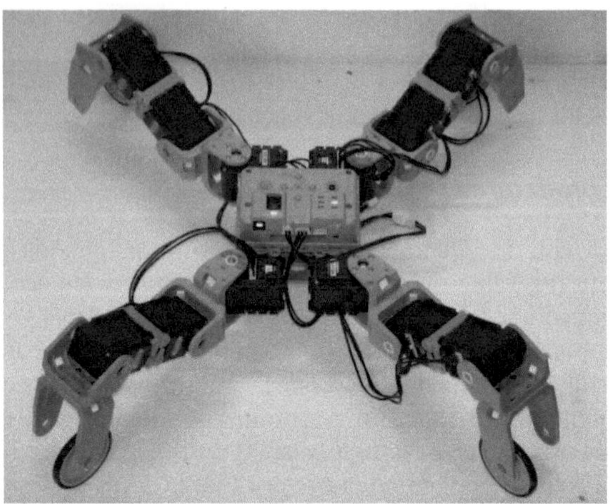

Fig. 4.3 Second design of the quadruped with a torso containing the battery

Nevertheless, the second design also failed immediately in performing a proper walking gait. It turned out that a design with long legs, as can be seen in the figure above, makes it almost impossible to ensure a safe tripod stand. This is because the center of gravity moves out of the imaginary triangle that can be drawn between the three points on the ground on which the robot stands when lifting one leg. More important, however, is the other disadvantage that the legs cannot curl up very closely to each other. This would be beneficial for the rolling position, since it reduces the diameter of a spherical structure that would later be assigned to the legs.

4.4.3　A Quadruped That Flips Over

A third robot was constructed by rearranging the actuators. An improved leg design is suggested to ensure a better walking performance and to allow flipping the quadruped over to the front, which also would be the first step in realizing the rolling locomotion. The outer actuators were now turned by 180° and mounted closer to the robot's torso with their joints. The wheels which were mounted on the tip of each leg before were replaced by aluminum prototype legs. So far, the wheels had no functional purpose, except providing some additional grip through their rubber O-rings when moving on slippery surfaces like on a laminate floor or on one with glazed tiles.

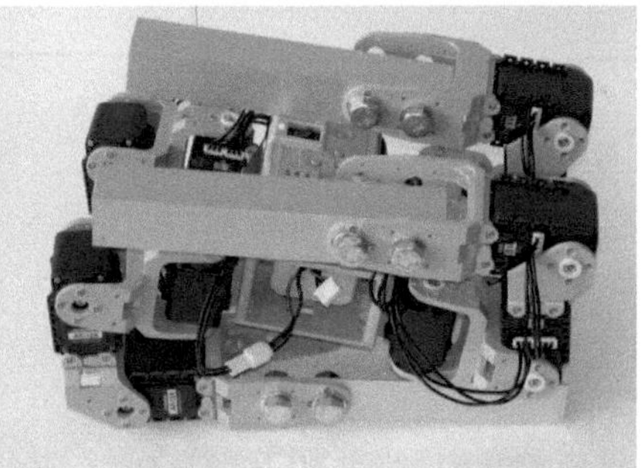

Fig. 4.4 A realignment of actuators allows the robot to curl up the aluminum prototype legs very closely

The proper length of the aluminum prototype legs is crucial to allow the robot flipping over. Shorter legs would have resulted in the robot pushing itself backward. Longer legs would have avoided flipping over as well, because they would have acted as a barrier once the robot reached the 90° position. To prevent malfunction of the servo motors due to an excessive torque that the motors would need to push the robot over, an L-shaped aluminum profile was considered to be the right component for a first prototype. Taking the weight of the fitting accessories into account as well, this prototype would have to prove if the robot can overcome the forces and torques of its own weight.

It appears like the alignment of the outer actuators differs between the front and rear legs, but in fact the robot has a symmetrical design when brought into the quadruped position. Taking the rolling position, however, will change the orientation of the joints, since one pair of legs went upward and the other pair of legs went downward. More symmetrical joint racks would have been necessary to realize a perfectly symmetric appearance in the rolling position as well. Unfortunately more joint racks were not provided with the robot kit and hence not available at the first testing stage. This circumstance results in a different length of the legs in the rolling position, which was not compensated by different lengths in the aluminum profiles.

It turned out during the first test phase that the influence of the small difference in the length of the legs did not noticeably affect the performance of the robot. Flipping over was achieved with the robot taking position (A) manually. Exceeding an angle of 90° resulted in an instant flip to the front (B and C). This is dependent on the speed of the actuators that can be set in the computer program. Slow pace requires a large angle to finally let the robot flip over like in section (C) of the figure. Fast pace would lead to a momentum in the robot that would let it flip over much sooner as can be seen in section (E) and (F).

Thereby the movement of the robot is already strongly inspired by the spider *Cebrennus villosus*. Kneeling on the first pair of legs, like the spider, to prepare for triggering the rolling locomotion is satisfyingly implemented into this first prototype. The torso of the robot does not touch the ground at any time, which is

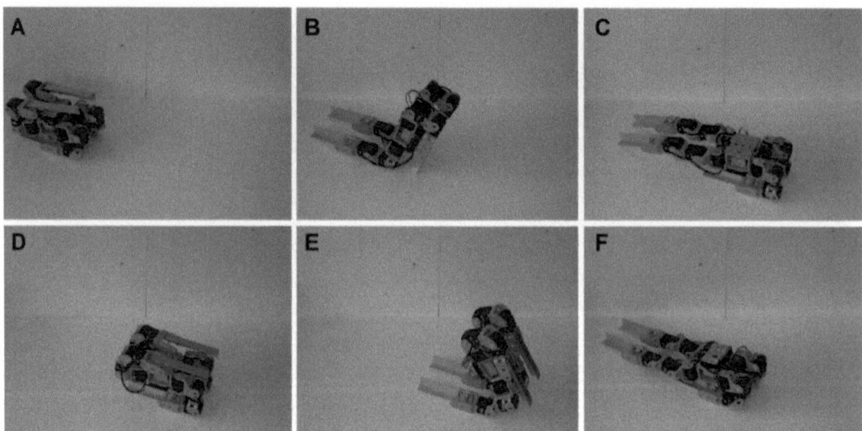

Fig. 4.5 Quadruped that flips over to the front

in line with the biological archetype. A huge difference is that the robot cannot perform a somersault in the air as the spider can. Actuation speed and maximum torque power, limit the robots capabilities.

A suggestion for further research after this project, concerning the robot, is that a somersault in the air might be achieved when the robot could reach a certain speed during the phase of rolling motion. Triggering a somersault with maximum actuation power and speed might hereby just be sufficient to lift the whole system into the air for the blink of an eye. The robot might then perform at least a 180° turn to the front, when it has a high revolving momentum. However this is just a theoretical assumption and surely requires adequate sensors with a high sampling rate.

So far, a movement of the robot was only achieved while it was connected to the computer. The software RoboPlus Motion, which is provided with the robot kit, is a very powerful tool to save motions of the robot quickly with a teach-in method. As mentioned earlier, the torque of the servo motors can be switched on and off individually. This allows moving the actuators into a desired position and assigning them to a single step in the motion program. These steps can be recalled again to let the robot take a certain position or perform a certain move. A motion sequence, containing up to seven steps, can be replayed nonstop. Whenever a single step of a motion sequence is recalled the robot always happens to move slowly. Replaying a whole sequence makes the robot move in real time again. The real time pace can be set in the RoboPlus Motion program directly for each single step and/or additionally for the entire sequence.

Changing a RoboPlus Task program or writing a new one from scratch was not necessary until now. However, for the implementation of a quadruped walking gait, it is necessary to replay more than one sequence, of which every single one contains a maximum of seven steps. This is because the walking gait is complex enough to require more than seven steps to move four legs in such manner that the robot will finally walk into the desired direction. Continuous sequences can only be replayed when called by a RoboPlus Task file. To exemplify the interaction of these two programs a simple quadruped was built following the manuals provided by the manufacturer ROBOTIS.

4.4.4 A Simple Quadruped to Exemplify Programming

The '(19) Quadruped Walking Robot' was chosen to provide a deeper insight and understanding of the interaction between the RoboPlus Motion and the RoboPlus Task program. Not all the robot examples provided by ROBOTIS need a RoboPlus Motion program, because the actuators can also be controlled with a RoboPlus Task program alone. Programming all the single actuators with just the RoboPlus Task program requires more effort than assigning the motions to steps and saving them as a motion sequence with the RoboPlus Motion program. This is obvious due to the fact that this option is only used with examples comprising very few actuators. However, the quadruped sample contains both - a RoboPlus Task and RoboPlus Motion file. Additionally this robot was chosen because it might give a hint to a good quadruped walking gate.

Fig. 4.6 Small quadruped example with software solution from ROBOTIS

The sample robot worked as desired and a closer look was taken on its programing and walking gait. In the following figure it can be seen how the command line, calling the motion file, was identified in the program.

The command lines for the forward locomotion of the quadruped can be seen in the RoboPlus Motion program highlighted in red. The English description allows fast orientation and differentiation among the command lines. Otherwise the command lines, responsible for the sought movement, can be found by connecting the robot to the computer and replaying the sequences in a single way. It becomes more obvious here why a more complex motion sequence needs to be called by the RoboPlus Task program. What can be seen is that in the line of 'Forward walk 1' the next line, which is called 'Forward walk 2', will be called with the line's number - here number three. Line three will then call line number two again, resulting in a continuous replay of the motion sequences that make the robot walk. In the RoboPlus Task program the function for calling the forward locomotion can

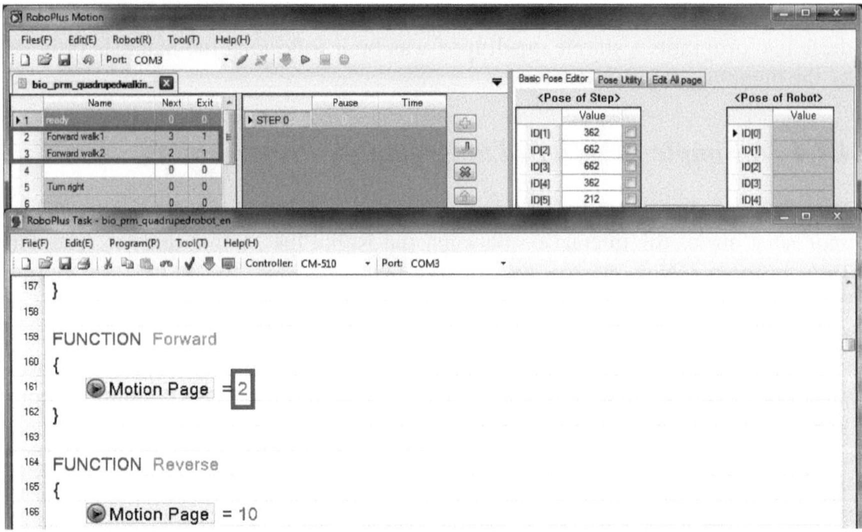

Fig. 4.7 Screenshot of the RoboPlus Motion and RoboPlus Task program identifying how a motion is called

easily be found with the number that would start the motion sequence. Knowing this, the number of the motion page can be changed and hence call any motion sequence created with the RoboPlus Motion program. This allows assigning any robot structure with any alignment of actuators to be controlled with one and the same RoboPlus Task file. But since there are no changes in the task file, there will be no changes in the behavior of the robot when approaching obstacles, for instance, either. The small quadruped example already contains a sensor for distance measurement and another one for object detection. This is beneficial for the quadruped mode, but disturbing for the rolling locomotion. Hence, it needs to be kept in mind that it will become necessary to change the RoboPlus Task program or to write a new one from scratch when a continuous rolling sequence is desired.

The walking gait of the small quadruped provides some interesting approaches. It works rather well for the small quadruped example and so the consequence is to try it out on the quadruped with aluminum legs, too. The program of the small quadruped only had to be changed in a way that the four additional actuators, which are necessary for curling up but not for walking, need to remain in either a straight or a 90° angled position. This was achieved by adding these actuators to the RoboPlus Motion file, which was downloaded from the ROBOTIS website, and assigning a fixed value to them. Already during the first two trials it turned out that the motion files then become unsuitable. The robot in its former appearance with the aluminum legs kept on dropping to the side and on the back while the forward motion sequence was played. The data therefore needed to be changed according to the specific values of each single actuator. The aluminum legs, which were great to let the robot flip over, then worsened the walking gait as

Table 4.2 Visualization of the new quadruped walking pattern, observed and changed based on the small quadruped example, now ensuring a safe tripod stand at any time

Move robot ⇨ Lift, lower and move leg ⇨	STEP 0 Move robot to the front	STEP 1 Lift one leg and move it to the front	STEP 2 Lower the lifted leg to the ground	STEP 3 Move robot to the front	STEP 4 Lift another leg and move it to the front	STEP 5 Lower the lifted leg to the ground
PHASE 01				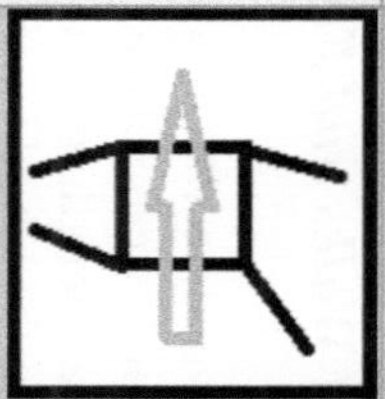		
PHASE 02		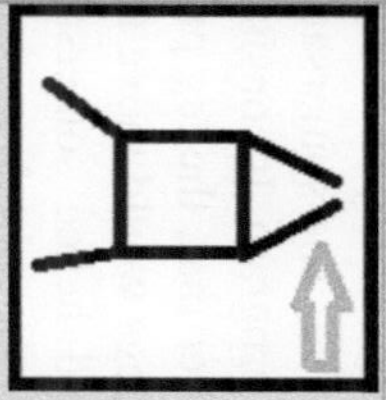	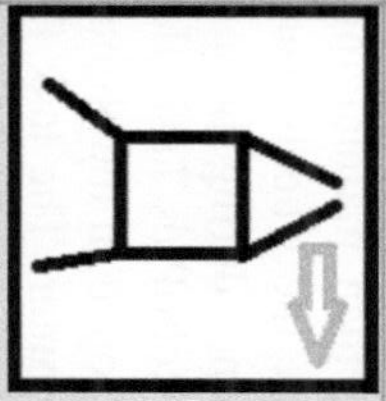	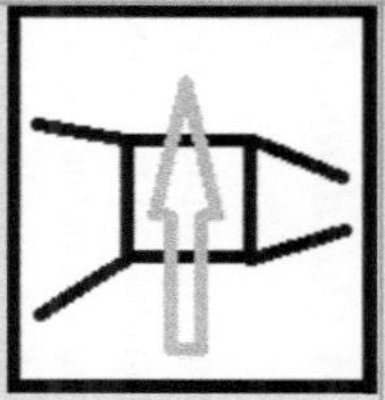		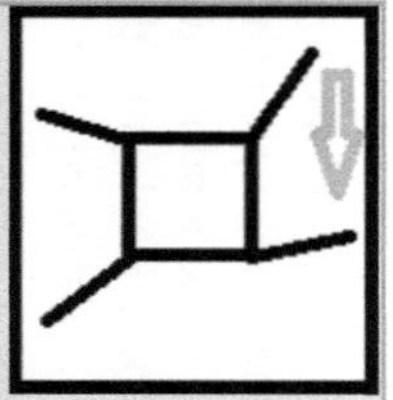

well. Especially the L-shape of the aluminum profile was not desired to provide a good ground contact when playing the walking sequence. Hence, a different leg structure had to be developed.

To find a principle pattern for a quadruped walking gait, a closer look has been taken at some random quadruped robot videos available online on the platform "YouTube". Replaying the motion file from the small quadruped step by step, however, provided the most useful insights. Throughout this testing and comparing, a movement pattern was extracted, which can be seen in the following figure. The sketches visualize a slightly different gait compared to that being performed by the small quadruped through the original motion file, which was provided by ROBOTIS. The visualized pattern differs to the original one by ensuring a safe tripod stand at any time. This was not necessary for the small quadruped sample, since it was suited with a good contact patch to the ground.

4.4.5 A Quadruped That Walks and Flips Over

To promptly implement and validate the new quadruped walking pattern, legs were mounted to the robot, which were supposed to fulfill the requirements of a good ground contact and a certain length to ensure somersault capabilities. Like in the very first quadruped version the wheels were used again, which have no other purpose than maintaining grip on slippery surfaces. In contrast to the former prototype the wheels were now turned 90° around the vertical axis to avoid any interference during the somersault motion.

The RoboPlus Task file was changed to call different motion files, which had to be developed in the RoboPlus Motion program with the robot being connected to the computer. The task file, which initially would call a motion to let the robot duck when an object is sensed above it, was now changed to trigger the somersault motion file. The somersault motion file had to be developed step by step with the motion program. It consists of a feature that lets the robot take a kneeling position first. After that, the leg pair in the back would curl up as well and finally the first pair of legs would move to let the robot flip over to the front. Flipping over to the front and hence somersaulting is programmed as an endless loop. With this program the robot's performance can be tested.

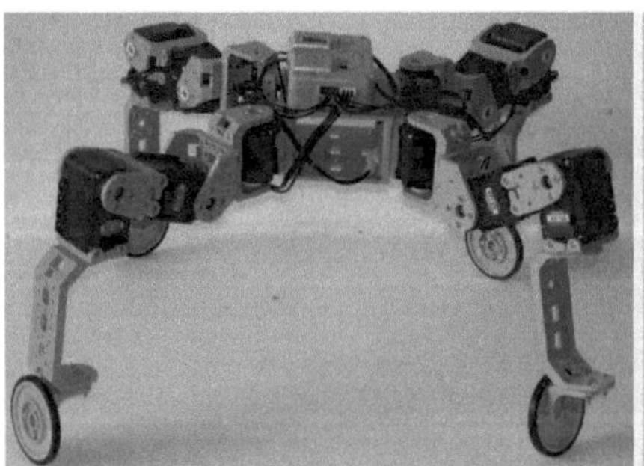

Fig. 4.8 Quadruped with a new leg structure for testing different movements

As recognized earlier, the walking gait with slight changes, especially in terms of the amount of actuators, was insufficient for the robot with aluminum legs. This turned out to be true for the advanced robot as well. While the kneeling and somersaulting motion were working perfectly fine, the walking mode still appeared to be an obstacle to be overcome. Motions created with the teach-in method, in which the torque can be switched on and off manually to save motion steps to a file, resulted in a somewhat retarded walking and crawling behavior. In some extreme cases the robot would even fall to the side and roll on its back to become entirely disabled then. Therefore the positions of each single actuator were taken from the motion file and transferred into a table similar to the following. The table consists of three phases. Phase zero represents the robot in its initial, standing position. Phase one and phase two, with six steps each, contain the position information of all actuators during the walking gait. Initially all values were incoherent in every step of the motion file. This is due to the reason that the servo motors cannot be placed absolutely accurate manually with the teach-in method. Hence, it became necessary to change all parameters to coherent values throughout the entire motion file. What can be seen in the following table is the optimized version of the walking gait.

Table 4.3 Optimized values of all servo motor actuators in the initial position and during the walking gait

ACTUATOR ID-NO.	PHASE 0	PHASE 1					
		STEP 0	1	2	3	4	5
1	362	**512**	512	512	**562**	**262**	262
2	662	**612**	612	612	**562**	562	562
3	662	**712**	**462**	462	**562**	562	562
4	362	**312**	312	312	**262**	262	262
5	512	512	512	512	512	512	512
6	512	512	512	512	512	512	512
7	512	512	512	512	512	512	512
8	512	512	512	512	512	512	512
9	212	212	212	212	212	**312**	**212**
10	812	812	812	812	812	812	812
11	712	712	**612**	**712**	712	712	712
12	312	312	312	312	312	312	312
		PHASE 2					

Table 4.3 (*continued*)

		STEP 0	1	2	3	4	5
1		**312**	312	312	**362**	362	362
2		**512**	512	512	**462**	**662**	662
3		**612**	612	612	**662**	662	662
4		**212**	**562**	562	**512**	512	512
5		512	512	512	512	512	512
6		512	512	512	512	512	512
7		512	512	512	512	512	512
8		512	512	512	512	512	512
9		212	212	212	212	212	212
10		812	812	812	812	**712**	**812**
11		712	712	712	712	712	712
12		312	**412**	**312**	312	312	312

Now, in the optimized table, a certain pattern for all actuators can be seen. It can also be seen that the actuators with the identification number five, six, seven and eight remain at the same value throughout all steps. This is to be explained by the fact that these actuators are necessary for positioning the legs in the rolling mode, but during the walking gait they can remain in their original position. The bold written values additionally highlighted in light green, exhibit a change in the values and thus a change in the actuators' angular position.

A comparison of Tab 4.2 and Tab 4.3 shows, that they perfectly complement one another. If the robot moved into forward direction, like in step zero of Tab 4.2, for instance, four values would need to change. Looking at step zero in Tab 4.3 it can be seen that in fact four values do change at the same time. Lifting a leg and moving it to the front at the same time requires a change of two values, because two DOF and hence two actuators are necessary for this particular action. This is coherent in Tab 4.3 step one and four, for example. The optimized values were chosen based on the fact that the value 512 resembles the initial position of a servo motor. Changing its position manually might result in a value like 608, for instance. Hence, the next higher number of 612 was chosen, which would be the result of adding a value of 100 to 512. For simplification of the motion values, it was decided to only add or subtract a value of either 50 or 100.

The reason for setting up a table like this is that it facilitates visualizing incoherence among the values during the single steps as opposed to taking a look at each single step one after another in the motion program. The optimized values gathered can now be reassigned to the values in the motion file, resulting in a very smooth movement of the robot. The robot needs to be disconnected from the computer to enable it playing the motion file in an endless loop. This is due to the reason that a loop among specific motion phases needs to be called by the microcontroller directly. Finally, a quadruped walking gait was satisfyingly achieved and performed by the robot. The robot shown in Fig 4.8 is now capable of walking. When an object like a human hand, for instance, is detected above the robot by its sensor, it will stop walking, take the kneeling position and then start to flip over to the front until switched off manually by the user.

After creating a structure with an effective alignment of actuators that is capable of walking and flipping over without a hardware malfunction, a leg design can be developed that enhances the robot's rolling capabilities. Until now, the robot lacks a continuous motion that needs to be achieved to approach the spiders' somersaulting technique.

4.4.6 BiLBIQ – The Rolling Quadruped

Before a set of prototype legs can be manufactured both their layout and dimension needs to be defined. The physical robot was therefore rebuilt in 3D with the CAD program "Autodesk Inventor Professional 2012"[3]. Especially the size of the legs can easily be defined using the program and a first leg design can be developed and tested on screen to unveil potential insufficiencies like a small movement range, for example, in advance.

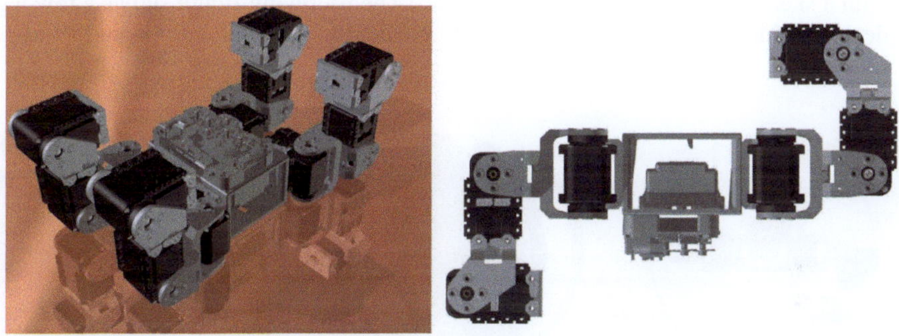

Fig. 4.9 Visualization of the rebuilt 3D robot model

The 3D data being used to rebuild the robot were taken from the ROBOTIS website: www.robotis.com. After having converted the downloaded files into a file format that could be used with the CAD program, different colors were assigned to the individual parts of the robot model. Especially the servo motors were colored with respect to the real robot to achieve a more realistic visualization. The individual parts were put together in a 3D assembly group by assigning position dependencies to them.

---

[3] "Autodesk" and "Inventor" are copyright protected trademarks of the company Autodesk Inc. USA.

The issue of having an irregular leg structure due to an incoherent alignment of actuators in the rolling position [and only in the rolling position, as mentioned earlier] can be seen in the right part of the figure shown above. Since an irregular design led to certain issues before, like avoiding a proper walking gait, for instance, it was decided to alter the leg design, even though the necessary additional parts for the realization are not provided with the robot kit.

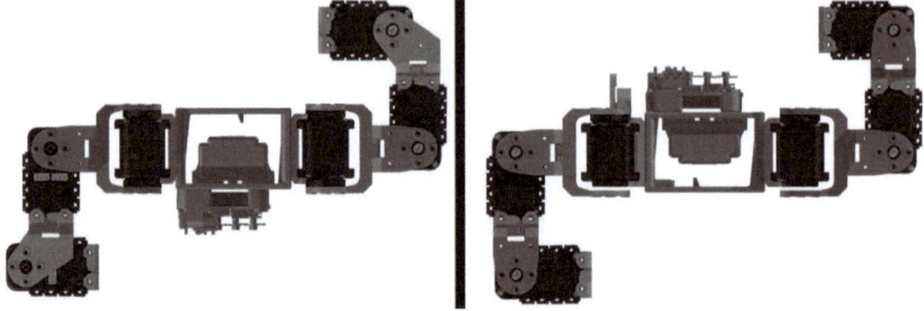

Fig. 4.10 Changing the outer mounting base on screen to achieve a more symmetric structure in the rolling position

Based upon this alteration a first approach was designed to achieve a structure similar to Fig 4.1 where a picture of the spider *Cebrennus villosus* was overlaid with a sketch of two circle-shaped 'wheels'. The following figure shows a raw leg design, which disregards the need of a suitable material and manufacturing process.

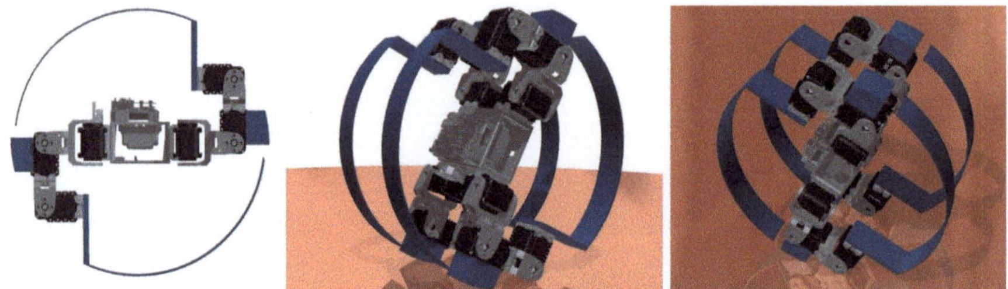

Fig. 4.11 First leg design thought to achieve a continuous rolling locomotion

With certain constraints and dependencies that were assigned to the 3D CAD model, such as limiting the movement range of the servo motors, a virtual testing of the configuration became possible. Another issue got unveiled when the kneeling, initial and walking position was simulated by altering the position of the individual actuators in the program.

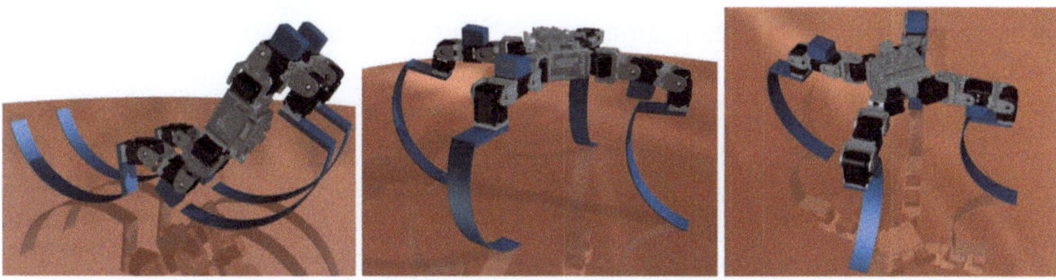

Fig. 4.12 Robot unfolding and resulting in an asymmetric initial and walking position

Now, when the robot would unfold to walk again, it could be observed that one pair of legs would point outward, while the other pair of legs would point inward. This circumstance is considered to be very obstructive for the walking gait due to the reason that the leg pair which points inward has now a very limited movement range. The actuators' alignment therefore needed to be reconsidered. Solutions in which some additional actuators would allow the legs to turn into a stable initial and walking position were evaluated. Finally, a leg alignment was found that would not only allow the robot to turn its legs into the desired position, but also supports some special poses of the robot like rolling sideward, for instance.

Fig. 4.13 New actuator alignment and new leg design allowing the robot to take entirely symmetric poses. For a better visualization the left pair of legs is displayed in blue color and the right pair of legs in red color.

As can be seen in the figure above, the amount of actuators has not changed. The difference lies within the orientation of the outer actuators on each leg that additionally do not longer require a mounting base like before, which anyway is not provided with the robot kit. In the virtual environment the robot is now able to take the walking and rolling position without any disturbances.

The design of the robot's legs is inspired by teeth, claws and shell-spines like they can be found in the book of Claus Mattheck[4]. The design principles that he developed, especially the use of his triangle method to allocate strain and stress throughout any structure like nature does, were a source of inspiration as well. However, it is refrained to follow the instructions accurately, because the relevance for the first prototype is considered as too low to spend too much time on this particular issue. The new leg design, in contrast to the first, now takes the

---

[4] Mattheck (2006).

process of its technical realization into account. It is suggested to realize the legs with a lightweight, durable and bendable material. Furthermore, it should be easy to assemble the individual leg parts. These requirements were met best by double-sided circuit boards usually used for the purpose of carrying electronic devices. The circuit boards that came to use here consist of an epoxy resin reinforced glass-fiber mat as core and are covered on both sides with a thin copper layer of about 0.035mm. The circuit boards come in two different layouts - one with a total thickness of 0.5mm - which is good for bending - and one with a thickness of 1.5mm - providing a better static stability. Advantageous about the circuit boards is that they can easily be soldered together with a normal soldering-iron and tin-solder. This bonding results in a very resilient and durable structure. Additionally the legs can be cut out of the board using a strong scissor, shortening the manufacturing time significantly.

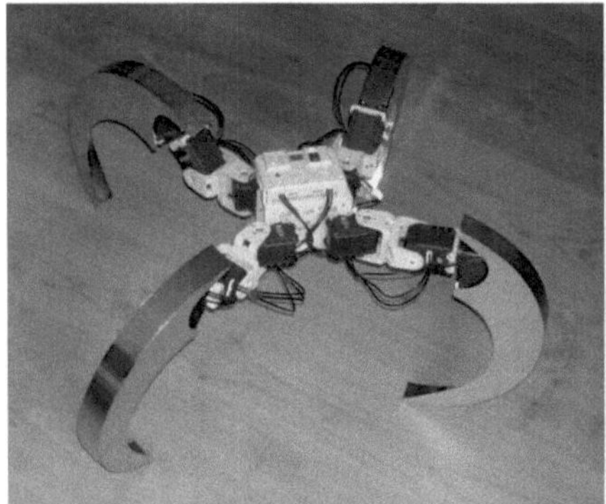

Fig. 4.14 Prototype legs mounted on the quadruped in standing position *(left)* and rolling mode *(right)*

The legs were finally manufactured by soldering the circuit boards together with a tinplate as tread that provides even more flexibility when bent. The legs' size and shape are based up on drawings which, in turn, were derived from the 3D CAD model. The robot's actuator alignment as well was changed referring to the virtual model that has proven its functionality before. After assembling all parts, the new prototype, as shown in the figure above, now needs to prove functionality in a real environment as well.

The programming needs to be changed now that different legs are mounted to the robot and especially because the last servo motors have a different orientation now as well. A new raw program was developed with the teach-in method like described earlier and optimized manually by adjusting the values in the motion program, similar to the method for developing a proper walking gait for the quadruped. The new program allows the robot to stand and take the kneeling position smoothly, which happens similar like in the quadruped from before, finally triggering the continuous rolling locomotion. To prevent the robot from rolling backward in kneeling position, the hind legs first need to move closer to the front and then curl up quickly resulting in a momentum to the front. When the robot remains in the rolling shape, as can be seen in the right hand part of the figure above, an actuation of the legs facing the ground will result in a somersault to the front. Since the robot becomes faster with each single actuation, the actuation speed needs to be increased as well. Propulsion works like this for up to three somersaults. Then the robot becomes too fast to tell in which way the program needs to be changed to trigger the respective pair of legs at the right time.

The walking gait could not be tested on this prototype since the legs turned out to be too bendable. First and foremost the robot tends to collapse when one leg was lifted as part of the walking process. The rolling locomotion, however, was performed satisfyingly. After curling up and triggering the propulsion motion three times, the robot kept on rolling for up to five meters on plain ground, as a result of the generated momentum. Finally it can be claimed that this first rolling prototype was a success, because rolling was achieved with a smooth continuous motion. It could be observed that the servo motor alignment works in principle in this new configuration. Moreover, this prototype shows where and in what way the leg design needs to be improved. In case of an entire failure of the design, it would have been necessary to reconsider it completely. Rather than that, the next prototype can profit from on the gained experience and is then expected to meet all requirements to effectively combine walking and rolling locomotion with one and the same structure.

Since the quadruped is expected to perform two very different kinds of locomotion it is therefore named: **"BiLBIQ the Bi-Locomotional Biomimetically Inspired Quadruped"**.

4.4.7   BiLBIQ 2 – Walking and Rolling

First of all the prototype robot BiLBIQ 1 posed the problem that the material thickness was chosen to thin to support the robots weight while standing on three legs. Additionally the shape of the legs contributed to weakening the structure as well. The legs were left open on one side to save material weight, but this strongly influenced the effect of distortion within the leg itself. Further considerations regarding the tread resulted in an enhancement of this structure as well to support better ground contact and a decreased probability of overturning while rolling. The next leg design should therefore take following aspects into account:

- Double-sided legs to avoid distortion.
- Thicker material to withstand the applied load.
- Increasing the size of the tread to prevent overturning.
- New design approach for a better distribution of loads.

4.4.7.1   Leg Optimization with FEM

The second leg design should prove its applicability before being manufactured for the robot. Thus the design approach is being confirmed in the CAD program with a brief FEM test. Again, the design is biomimetically inspired only and following exactly biological principals needs to be disclaimed. A biological

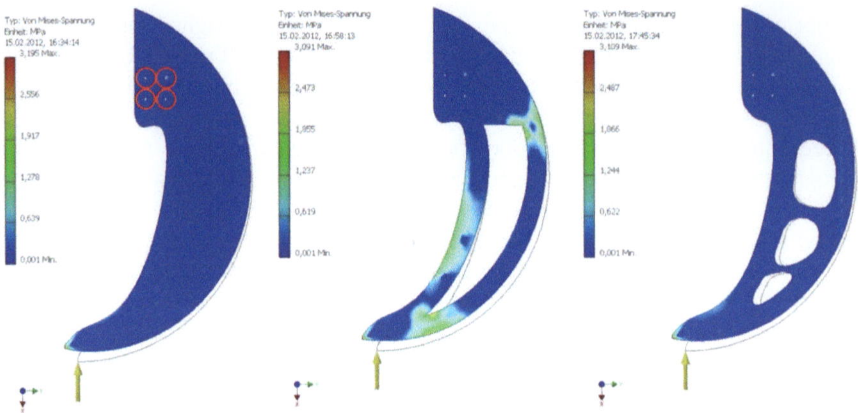

Fig. 4.15 FEM analysis of the biomimetically inspired leg design

principle that was tried to be obeyed, however, is referring to a statement of Mattheck, which says that the part of a structure that needs to withstand higher loads, needs in return more space and hence more material[5].

The figure above displays the design approach and its progress. While the leg in the left part of the figure is only shaped in a way to not collide with the robot's body and especially the microcontroller when curled up, the leg in the middle part, in contrast, was designed to save material and thus weight. Here it can be observed how the load is distributed unequally throughout the structure.

For the FEM analysis the structure was fixed in four points - highlighted with four red circles in the left part of the figure. Creating a fixing point in the virtual environment means that the structure cannot rotate around or move along an axis in this point [and in this point only], thus having zero DOF. The reason to make these holes the fixing points is that the four points are needed to mount the servo motor later. The load, indicated with a yellow arrow, is applied on the tip of each leg, like it would be the case when the robot remains in standing or walking position.

The load value was derived from the estimated weight that the leg would need to carry when one of the legs is lifted for walking. Following suggestions were made to derive the applied load: The robot has a total weight of 1114g without legs. The legs of BiLBIQ 1 have a weight of 56g each. The new legs are expected to be three times heavier, because they are double sided, having a higher material thickness and a larger tread. Four leg's with a total weight of 672g plus the robot's weight result in a total weight of 1786g. If one leg is lifted for walking the legs remaining on the ground would each need to withstand one third of the robots total weight, hence roughly 600g. The 600g load per leg needs to be divided into two, because one leg consists of two of the structures tested with the FEM analysis. 300g plus a small safety margin of one fourth result in a total weight of 400g. Since 100g weight approximately equate the force of one Newton, the applied load can be defined with four Newton. Each of the structures displayed in Fig 4.15 were therefore loaded with the force of four Newton.

---

[5] cf. Mattheck (2006) p 45.

One tradeoff regarding the testing of the material in the FEM analysis had to be made. Since circuit boards, or epoxy resin reinforced glass-fibers are not listed in the material list of the FEM analysis that was used, a polycarbonate with similar assumed characteristics like stiffness, for instance, was chosen instead. Since the real characteristics cannot be simulated, a test until failure of the structure was not run through the FEM analyses. Nevertheless, for demonstrating the approach of an FEM analysis the computer model and simulation was regarded to be sufficient.

By obeying the 'more load - more material principle' mentioned earlier in the this chapter, the opening was reinforced with two bars and the resulting corners were smoothened out with biologically appearing curvatures as it can be seen in the right part of Fig 4.15. Again the applied load now distributes coherently throughout the entire structure like in the first design on the left hand side of the figure - but now with the advantage of less material usage.

4.4.7.2 Assembling the Robot Using CAD

Again, it is applicable to assemble the robot with the CAD program first to uncover hidden problems in the new design, before manufacturing it. The finalized leg structure now comprises two of the FEM tested leg designs and the tread, which in contrast to BiLBIQ 1 is soldered from the outside of the structure for convenience reasons.

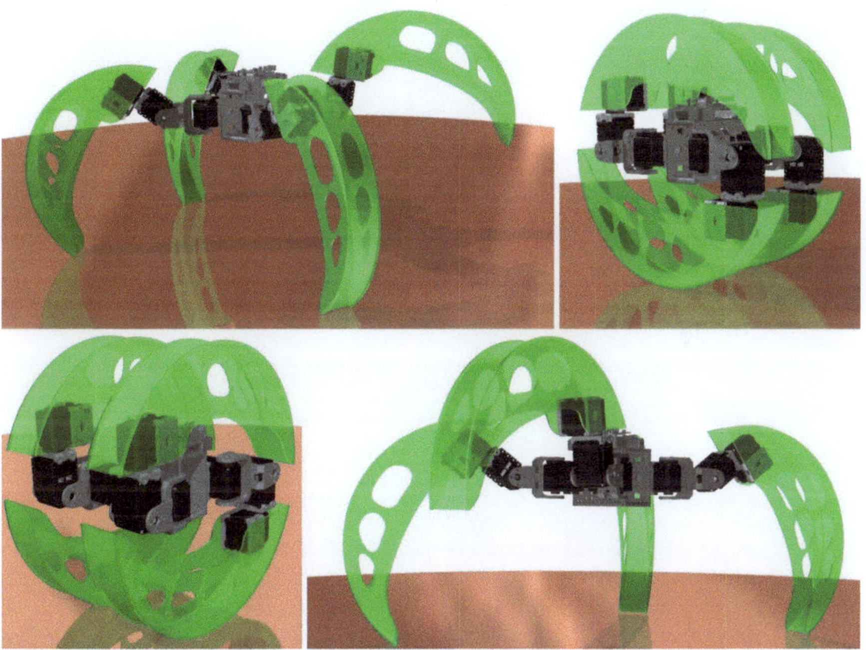

Fig. 4.16 BiLBIQ 2 in rolling and walking position with one leg lifted. The legs are indicated in transparent green color for a better contrast in the CAD renderings.

The new leg design takes care of the considered aspects mentioned before. As displayed with the FEM analyses the load distribution remains similar even though there are three cut-outs in the leg to save material and hence weight. The legs are now double-sided to mainly prevent distortion. The size of the tread is increased in width in consideration of ensuring a safe condition during the rolling locomotion. In terms of material choice it was decided to stick to circuit boards. The relation of weight to righty and cost is very good compared to other materials like carbon fiber-mat, for instance, which additionally require a different manufacturing and joining process.

4.4.7.3 Manufacturing the Legs and Assembling the Robot

The CAD data was converted for a circuit board milling-machine that was used to cut out the final shape of the legs. The side elements of the leg have a thickness of 1.5mm each, in contrast to the first design with a thickness of .5mm only. The tread's thickness, however, still measures .5mm, because it needs to be bent along the legs edge. This process would not have been possible with 1.5mm strong circuit boards. As mentioned earlier the legs were then soldered from the outside

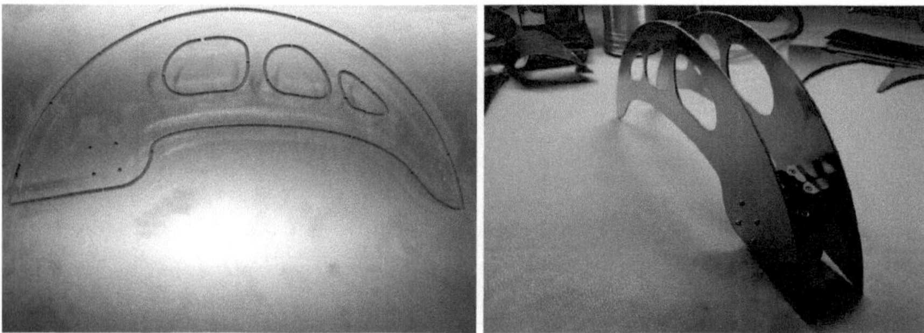

Fig. 4.17 Leg cut-out by the milling-machine *(left)*. Mounting two parts on a servo motor for soldering *(right)*.

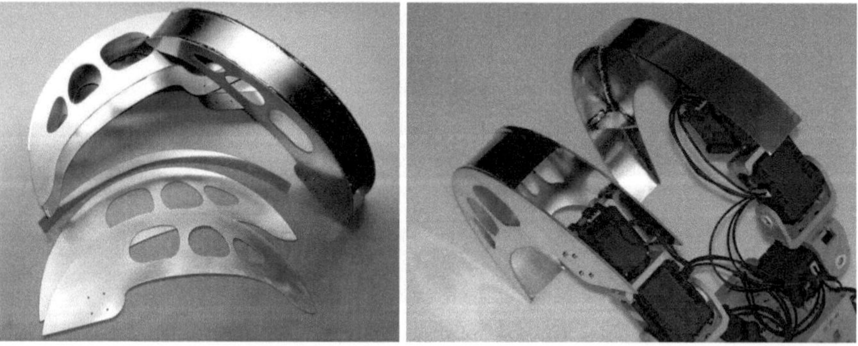

Fig. 4.18 Single leg parts and soldered legs *(left)*. Comparison of leg version 1&2 mounted on the robot *(right)*.

along the tread for reasons of convenience. But the robot's performance, however, is thereby not affected in a negative way.

As can be seen on the right hand side of the figure above, the tread of the new leg design is wider than the one of the previous version. It also does not cover the entire leg anymore as to allow an easy access to the cables' sockets. Since the tread in fact is not used as such, because the robot is supposed to primarily roll on the edge of the legs, a negative effect on the rolling motion is not to be expected. Distortion and similar unwanted effects on this part of the leg, which appears to be weakened at the first glance, are avoided as well due to the reason that both side elements are mounted to the servo motor bases' left and right hand side.

With 114g each the weight of the new legs has almost doubled in comparison to the previous version with a weight of 56g each. Nevertheless the stability exceeds all expectations and cannot be compared to the previous version. Finally the rolling locomotion and a walking gait can be implemented with this new legs mounted to the robot. First of all I had to test whether the increased weight of the robot causes a hardware failure, or if any other negative side effect occurs.

4.4.7.4 Testing the New Design

Since the actuator alignment of BiLBIQ 1&2 is the same and the size of the circle shape did not change either, the program written for the rolling locomotion can be used again. First of all I tested the rolling capabilities with the robot in its circular shape as initial position. Upon triggering the pair of legs that faced the ground, the robot instantly moved, resulting in a forward somersault motion. Therefore it can be derived that the robot is still light enough to overcome gravity forces.

For a better demonstration of the single steps that result from playing the motion program, a slow and thus stable behavior is sought in which the robot changes from standing to rolling. The raw programming for that was done again with the teach-in method. An adjustment of the individual values of each servo motor completed the programming. While a changed version of a given task program was used before, it was decided then to write a very short code in the RoboPlus Task program. The program only contains one endless loop in which different motion files can be called through pushing one of the four buttons on the microcontroller. This is very useful for testing and demonstrating different motions without the necessity to change the program in the microcontroller and thus connecting and disconnecting it from the computer.

The difficulty in programming the robot to perform the movements in slow motion lies with the fact that no momentum, resulting from a fast movement of an actuator, can be used. Through testing it was found that curling up the robot at a fast pace from standing to rolling position results in an instant rolling locomotion without the necessity to trigger the legs explicitly for propulsion. Even though this behavior reminds of the spider, which curls up very quick as well and seems to use this momentum, a final statement based on this assumption cannot be made without further investigation of the biological sample.

Nevertheless it was managed to program and play the robots motion at a slow pace with the tradeoff that the hind legs move one after another and thus in an asynchrony manner. This is not coherent with what the spider does, but the chance to control, observe and adjust the robots movements in a stable condition were more important than achieving results only by trial and error. Especially the fact that the robot does not bear any sensors up to now makes it important to control all states and movements as far as possible and to keep them in a stable condition. The rolling locomotion, of course, cannot be considered to remain stable. The fact that the robot needs a sensor for recognizing its angular position to trigger the respective legs for propulsion at the right time can only be suggested by now, because the implementation of such a sensor would exceed the scope of this book.

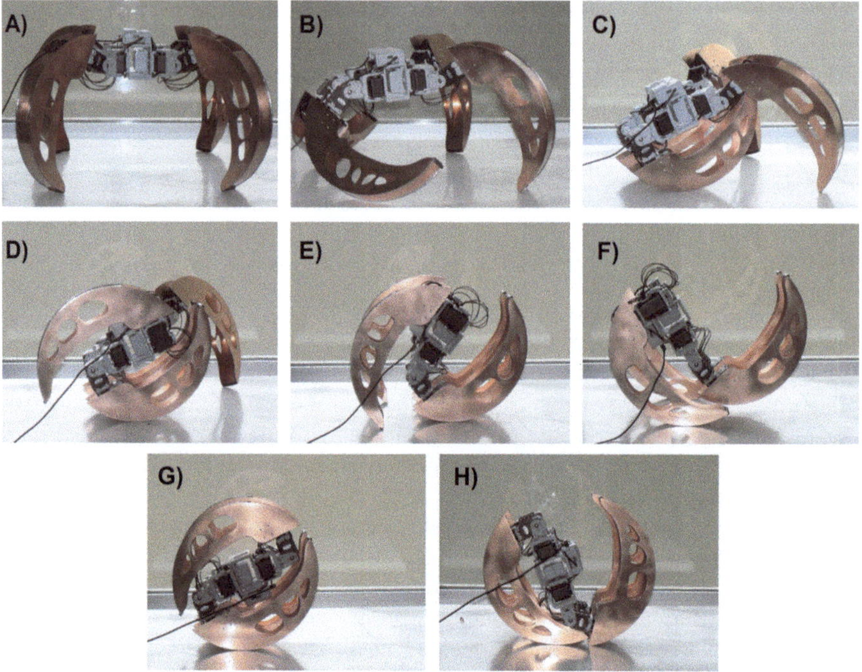

Fig. 4.19 BilBIQ 2 switching from standing position to rolling condition to trigger the somersault motion

The movement-activity as seen in the figure above can briefly be described as follows: The robot's initial position is the standing condition (A). The robot starts the movement by kneeling on the front pair of legs (B). Now the front legs are positioned below the robots torso thus resembling their final rolling position (C). The left hind leg moves to the front and places itself in a position that supports rolling (D). After that, the right hind leg can move to the front as well (E). Along with the right leg the gravity center of the robot moves to the front. The left hind leg, which is already positioned in the front, prevents the robot from instantly flipping over while the right hind leg is still moving. Both hind legs are now close to their final rolling position. To trigger the rolling locomotion, the front legs, which are still below the robot, actuate to push the robot over to the front onto the hind legs (F). The robot starts rolling. Before the hind legs move to their final rolling position, they trigger another somersault with an actuation movement directed at the ground the legs are facing at that moment. The robot performs propulsion by alternatingly actuating the respective pair of legs facing the ground.

In theory each actuation results in an even faster somersault with every subsequent turn. Therefore the actuation speed needs to be increased as well. In this case each actuation in fact does result in an increasing somersault pace. I managed to trigger the respective pair of legs facing the ground for four times. Hence, the front and hind legs alternatingly actuate for two times each. Already then, the speed increases so much that it becomes hard to tell which legs need to be triggered next by merely observing the movement.

BiLBIQ 2's walking gait is basically achieved like the one in the quadruped where the four external actuators had a different orientation. Due to the fact that four of the actuators could remain in the same position during the entire walking gait tested on the quadruped sample before BiLBIQ 1 was built, the alteration of the actuators' alignment did not significantly influence the new gait. The task of lifting the leg is now carried out by the inner servo motors which remained at a fixed angular position before. The new leg design, however, required some changes to the values of the motion program. The center of gravity in relation to the mass of each leg is now harder to control since the legs are heavier compared to those of BiLBIQ 1. Hence, the robot needs to take smaller steps to prevent losing its balance.

Fig. 4.20 BiLBIQ 2

References

Armour, R.H.: A Biologically Inspired Jumping and Rolling Robot. Dissertation University of Bath (2010)

Geheorghe, V., Alexandrescu, N., Duminică, D.: Design and experimental research on the displacement of an original construction of rolling robot. U.P.B. Sci. Bull., Series D 73(2) (2011),
`http://www.scientificbulletin.upb.ro/rev_docs_arhiva/` `full29700.pdf` (accessed February 19, 2012)

Mattheck, C.: Verborgene Gestaltgesetze der Natur: Optimalformen ohne Computer. Forschungszentrum Karlsruhe GmbH, Karlsruhe (2006)

Rechenberg, I.: Personal Interview (January 27, 2012)

`http://robot-kits.org/` (accessed February 13, 2012)

`http://www.bionik.tu-berlin.de` (accessed January 06, 2012)

5 Results

5.1 The Robot's Performance

The latest robot prototype, named BiLBIQ 2, is able to perform a walking gait and a rolling locomotion. The programming that is implemented so far, allows the robot to take the rolling condition from a standing position in a stable, hence controllable, manner. The movement the robot performs to take the rolling position and the rolling locomotion itself is strongly inspired by its biological archetype – the *Cebrennus villosus*. The robot obeys the biomimetic principles that were derived from the research material accessible and the information gained from the interview with Ingo Rechenberg. Nevertheless it needs to be differentiated clearly that BiLBIQ 2 is not a biomimetic robot but a biomimetically inspired implementation. This is due to the reason that an accurate biomimetic implementation would require more research on the biological sample to exactly understand the whole activity chain performed by the spider.

BiLBIQ 2 has a total weight of 1570g including the battery pack. With the current walking gait and actuation-speed it is capable of walking 0.06 m/s. The rolling locomotion is roughly twelve times faster with a speed of 0.7 m/s. Both walking gait and rolling locomotion have not yet reached an optimal state. The actuation speed of the servo motors is not yet maximized either. Therefore it can be claimed that there is still potential regarding speed, which was defined as an efficiency parameter in chapter 2.1.2. An optimum can be reached with programming change. This requires more development, preferably along with the implementation of sensors to the robot. Especially the robots rolling locomotion is expected to enhance significantly along with the usage of an angular measurement sensor. This sensor could provide feedback on the current position of the robot and thus trigger a propulsion actuation at the right time.

The robot is not yet able to keep up with the spider *Cebrennus villosus* even though, with a width of 600mm when standing, it is six times bigger than the animal [see Tab 2.1]. Without taking scaling effects into account the spider is still roughly three times faster in rolling and sixteen times faster in walking condition.

Nevertheless the main purpose of combining two fundamentally different ways of locomotion was successfully implemented into a robot prototype that was developed without using additional, unnecessary components. The legs designed for rolling and the alignment of actuators do not prevent the robot's ability to walk. Moreover they still allow the robot to roll. In addition to that it can be said that the rolling locomotion resembles an effective enhancement of the robot's

Ralf Simon King: BiLBIQ: A Biologically Inspired Robot, BIOSYSROB 2, pp. 77–82.
DOI: 10.1007/ 978-3-642-34682-8_5   © Springer-Verlag Berlin Heidelberg 2013

abilities since the rolling locomotion is twelve times faster than the walking gait. As a result the tenor question from the beginning can be changed into a statement:

With the robotic implementation of BiLBIQ 2 it could be shown that it is possible to mimic two fundamentally different ways of locomotion with one and the same structure – effectively!

5.2 Field of Application

A robot like BiLBIQ 2 can basically be deployed in every place where a combination of good terrain handling with rapid movement on suitable surfaces is sought for. The robot could have reconnaissance purposes when used as a mobile science or sensory station. If a data grid should be erected, a bunch of robots could be released in one spot of the area under investigation to let them reach their desired destination autonomously. Thus it would not be necessary to set up several operational science stations one by one. Especially in contaminated areas, temporarily or permanently harmful to humans, the robot could collect data. In contrast to an immobile sensor station the robot is able to change its position at any time for a higher accuracy and precision of the data that is collected.

Desolated and contaminated urban areas would be an ideal operational area for BiLBIQ 2. The presence of a partially intact infrastructure like streets or pathways, for instance, can effectively be used for a rapid locomotion of the robot with its rolling capabilities. If an obstacle occurs the robot would be able to circumvent it with his walking abilities to cross terrain where rolling is not possible. A very safe position could be achieved when the legs are spread and finally the investigation like an analysis of ground samples, for instance, could begin.

Another field of application would be supporting firefighters and rescue teams with areal data about toxicity of contaminated air caused by a chemical accident, for instance, or information about the spreading of fire sources in a burning

Fig. 5.1 BiLBIQ 2 *(right bottom corner)* investigating a desolated area with partially intact infrastructure

Fig. 5.2 Example of how BiLBIQ 2 could climb an obstacle when programmed accordingly

Fig. 5.3 Robot taking a safe position to start investigation on ground samples

refinery or similar structure. If the situation would become literally too hot for the robot it is able to rescue itself and its [cost-intensive] sensory, like the spider which escapes from a predator by curling up its body and starting to somersault.

A further field of application could be the investigation of canal systems. Long canals could be investigated to find disturbances like bulky waste that got stuck somewhere or leaks in the walls and pipes, for instance. Since the robot is able to roll at high velocities it could overcome long distances in a short time. If an obstacle occurs it would unfold and climb it in its walking gait, reporting or saving the position of the congestion to be removed with suitable equipment.

Major advantages of BiLBIQ 2, which merges rolling and walking locomotion in the same structure, are: That there is no additional structure like wheels, for instance, which are of no use during the walking gait; That fast rolling can be achieved with low actuation movements but without parts that need to revolve continuously and that thus the efficiency in terms of energy consumption can be kept low; That a body structure which is able to roll on plain surfaces can roll down slopes as a result of gravity force without any actuation and thus without any energy consumption at all; That, since the whole body serves as a tread with a large diameter, the robot can roll down slopes and overcome obstacles in its way without unfolding, whereas a wheeled robot would fail to overcome them.

5.3 Conclusion

5.3.1 Summary

The book 'BiLBIQ: A biologically inspired Robot with walking and rolling locomotion' deals with implementing a locomotion behavior observed in the biological archetype *Cebrennus villosus* to a robot prototype whose structural design needs to be developed.

The biological sample is investigated as far as possible and compared to other evolutional solutions within the framework of nature's inventions. Current achievements in robotics are examined and evaluated for their relation and relevance to the robot prototype in question. An overview of what is state of the art in actuation ensures the choice of the hardware available and most suitable for this project. Through a constant consideration of the achievement of two fundamentally different ways of locomotion with one and the same structure, a robot design is developed and constructed taking hardware constraints into account. The development of a special leg structure that needs to resemble and replace body elements of the biological archetype is a special challenge to be dealt with. Programming is a basic necessity to judge the performance of the robot. Finally a robot prototype was achieved, which is able to walk and roll - inspired by the spider *Cebrennus villosus*. Yet the necessity of implementing sensors to the structure becomes obvious and is a matter of further research and development.

5.4 Outlook

5.4.1 Programming

A robot with a certain amount of servo motors needs to be controlled by a microcontroller. This is due to the reason that a person's abilities to control the actuators of a complex system are limited. With a radio-control, for instance, a toy car, tank, or even a helicopter model can be controlled [it needs to be mentioned here that the new toy helicopters already contain sensors and feedback control loops that enhance the controllability significantly]. But a robot like the one that was developed during this project exceeds the average user´s abilities to control every state. Hence programming is the basic necessity for bringing 'life' to the structure.

In case of BiLBIQ 2 it was managed to achieve locomotion with the programs RoboPlus Task and RoboPlus Motion that were included in the robot kit. Even though the motion program contains the powerful teach-in method, limits are soon reached when it comes to implementing own sensors to the system. Therefore it needs to be mentioned here that programming a future approach might require different means like programming the robot in C language, for instance. This is supported by the manufacturers program.

5.4.2 Sensor Implementation

The locomotion performance of BiLBIQ 2 is mostly limited by a lack of sensors. The robot's structure holds many capabilities like reaching a maximum speed in the rolling condition, for example. But to achieve this, at least one sensor with a high sampling rate is required that continuously gives feedback about the robots angular position. Furthermore a behavior or at least an interaction with the environment could be achieved with the use of additional sensors such as distance measurement, light detection and tactile sensors. Allowedly in terms of control and 'intelligence', achieved with sensors and certain programming, BiLBIQ 2 is still in its infancy.

5.4.3 Leg Design

The relevance of a suitable leg design, regarding the configuration of the servo motors as well as the shape of the legs itself, was exemplified throughout the entire book with several prototypes. Even the robot of the project before this book showed the major significance of a proper leg structure. Material choice and manufacturing methods needed to be evaluated. Moreover, the development of the leg's shape itself needed to be considered thoroughly. Classical static calculations do not represent state of the art approaches. Dealing with biomimetic optimization requires more than what was displayed within this book. Nevertheless, an example of what is possible with the FEM analyses was provided with the leg design approach of BiLBIQ 2.

5.4.4 BiLBIQ or Hexapod-Sphere?

In chapter 4.4.1 the option was considered of developing the biomimetic complement to the spider as either a quadruped or a hexapod. Even though the approach of a quadruped is reasonable and the implementation was a success, as demonstrated in this book, the mind should be kept open for the possibility to realize the biomimetically inspired robot with a different leg configuration. A hexapod might offer advantages that were not foreseen in this book. A concrete idea for a hexapod could be to combine it with a Reuleaux-triangle shape instead a circular one, like Prof. Ingo Rechenberg did.

5.4.5 *Further Approaches*

Entirely different approaches can be thought of if time and budget constraints are of secondary relevance. The robot's performance in terms of actuation speed in this project strictly depends on the characteristics of the given hardware. In case of a free hardware choice a robot could be built that might use actuators that are only developed for its purposes and needs. Hence the performance in terms of fast locomotion as a result of a high actuation speed could be significantly enhanced. Hardware independent approaches might be the usage of different actuation methods like presented in chapter 3.2. The chance to obey more of the spider's locomotion principles to achieve a robot prototype relying on biomimetic findings more strictly is another advantage if the hardware could be designed freely. However, more research on the spider *Cebrennus villosus* needs to be done first, before this is going to be thought about too much.

Personal Resume

In my thesis titled 'Biomimetic inspired Robot Design with Walking and Rolling locomotion', which allowed me to achieve the degree 'Master of Science' and on which this book is based upon, I was able to incorporate very different fields of knowledge. I dealt with the biological sample for my project, the spider *Cebrennus villosus*, even though I was not given the opportunity to examine the spider in reality. Knowledge was gained about its anatomy and how the spider lives and 'functions'. The spider was compared to other animals and some framework requirements were concluded to be obeyed in the further proceedings of my work.

Different approaches were considered, taking state of the art technology into account. Limits were set by given hardware constraints and required special attention while developing the robot prototype. A structural design, inspired by nature, was developed and enhanced throughout the thesis. Programming became crucial to be capable of achieving any kind of positive result with the robot.

Finally I claim to have successfully developed a biomimetically inspired robot, obeying the main principles of its biological archetype, the spider *Cebrennus villosus*, as far as possible, based upon all knowledge that was accessible to me at the time of my research. With this book I hope to inspire researchers working in adjacent fields of science and I wish that the results of my work will be used as a basis for ongoing and novel projects in mobile, biomimetically inspired, robotic design.

If you have any concerns about our products,
you can contact us on
ProductSafety@springernature.com

In case Publisher is established outside the EU,
the EU authorized representative is:
Springer Nature Customer Service Center GmbH
Europaplatz 3, 69115 Heidelberg, Germany

Printed by Libri Plureos GmbH
in Hamburg, Germany